THE PSYCHOANALYSIS OF SCIENCE

This book is devoted to the memory
of my dear friend
Prof. Amos Funkenstein

THE PSYCHOANALYSIS OF SCIENCE

The Role of Metaphor, Paraprax, Lacunae and Myth

YEHOYAKIM STEIN

sussex
ACADEMIC
PRESS

BRIGHTON • PORTLAND

2 4 6 8 10 9 7 5 3 1

First published 2005 in Great Britain by
SUSSEX ACADEMIC PRESS
Box 2950
Brighton BN2 5SP

and in the United States of America by
SUSSEX ACADEMIC PRESS
920 NE 58th Ave   Suite 300
Portland, Oregon 97213-3786

British Library Cataloguing in Publication Data
A CIP catalogue record for this book is available from the British Library.

Library of Congress Cataloging-in-Publication Data
Stein, Yehoyakim.
    The psychoanalysis of science : the role of metaphor, paraplax,
lacunae, and myth / Yehoyakim Stein.
        p. cm.
    Includes bibliographical references (p.   ) and index.
    ISBN 1-84519-070-X (hardcover : alk. paper)
    1. Science—Philosophy. 2. Psychoanalysis and philosophy.
    I. Title.

Q175.S7675 2005
501—dc22
                                                            2004026127
                                                            CIP

Typeset & Designed by G&G Editorial, Brighton
Printed by The Cromwell Press, Trowbridge, Wiltshire
This book is printed on acid-free paper.

Contents

< Contents >

Preface

If, as Arthur Koestler comments in *The Sleepwalkers*, his book on the development of astronomy, the history of science is a relative newcomer on the intellectual scene, then the psychoanalysis of science can hardly be said to exist at all. For, while the works of Thomas Kuhn, Karl Popper, Paul Feyerabend and Imre Lakatos have become canonical in the history of science, research relating to the psychology of science has been scattered and sporadic. Koestler tacitly makes this point when he notes that while the progress of science is generally regarded as a kind of clear, rational advance along a straight, ascending line, in fact its path has been much more of a zigzag. The history of cosmic theories, in particular, he says, may be called a history of collective obsessions and controlled schizophrenias, and the manner in which some of the most important individual discoveries were arrived at reminds one more of a sleepwalker's performance than that of an electronic brain.

Processes of creativity cannot, in my opinion, be explained only in terms of psychopathology, but there is no doubt that the localization of psychopathology in developmental processes critically widens the scope of science. This is something that the history of science has not truly absorbed. This book, therefore, deals with the psychoanalysis of science—which is to say, it offers an analysis of the irrational elements involved in scientific development. I try to show how blind spots in science have hindered scientific development, and how identification of these blind spots can enhance scientific creativity.

Ideally, hidden currents and unconscious motives in science should be studied in an interdisciplinary way, combining history, philosophy, sociology, and psychoanalysis. Those currents and motives are not to be explored and explained solely on the basis of the personal psychology of the scientist (the more conventional approach); they need to be viewed as part of the larger development of science. They are phenomena produced by the scientific community as a whole. Systematically applied to all branches of science, then, the psychoanalysis of science can contribute to the awareness of the irrational motives that affect scientific progress. The detection of unconscious

< PREFACE >

blind spots will widen the horizon of science and lead to new discoveries.

In other words, while the science on which my research is applied in this book is psychoanalysis itself, my hypothesis is that the method of analysis that I apply can be used in other disciplines as well. Much as my study started out with the search for some missing links in the theory of dreams, and gradually widened my interest in the matter, so a study of the psychoanalysis of psychoanalysis would ideally branch out into the psychoanalysis of other scientific fields.

This book is based on lectures and seminars I gave during recent years in the department of History and the Philosophy of Science at the University of Tel Aviv, and in the Israel Psychoanalytic Society. I would like to take this opportunity to acknowledge the support of several people who helped me develop this material into its present form.

First, I am deeply indebted to my friend and colleague Professor Amos Funkenstein, who encouraged me at every step in this project, always with intellectual penetration and heartfelt comments. I am also deeply indebted to other friends and colleagues who helped me during different stages of my work. In particular I would like to express my thanks to Professor Heinz Stephan Herzka and Professor Bennett Simon, who discussed parts of the book with me and inspired me with their scientific perspective. The late Dr Ruth Zafrir and Dr Rami Aronzon provided constant support and lent me, as it were, a third ear.

I would like to express my gratitude to my children Arbel and Alma, and my mate Dina – they are a wonderful light in my life.

< Preface >

PART I

Locating the Irrational, Understanding the Repressed

Part I provides an outline of the pyschoanalysis of science. Various concepts relating to the unconscious in science are introduced and the rationale for the psychoanalysis of the discipline itself is demonstrated. The argument presented is that the rationale is derived from the existence of subjective and irrational elements in science. Philosophical claims are made for the discipline's subjectivity.

Genres of the historical research that deals with the unconscious are discussed, and the tools with which I seek to examine the scientific unconscious are described and explained.

Part I concludes with several archeological analyses of psychoanalytic historiography, close in spirit to the mode of thinking and analysis presented in Parts I–V.

Why a Psychoanalysis of Science?

Motto: the studies that follow are studies of history by reason of the Domain they deal with and the references they appeal to; but they are not the work of an historian. The object was to learn to what extent the effort to think one's own history can free thought from what it silently thinks, and enable it to think differently. — *Michel Foucault*

The *idea* of a psychoanalysis of science was first raised by Gaston Bachelard in a series of books he published between 1927 and 1953. These books refuted an *a priori* philosophical idea of rationality and suggested, instead, that philosophy should be grounded in science. Science, as Bachelard saw it, should form the cornerstone of philosophy.[1]

Bachelard, however, did not believe that science progressed in a linear fashion. In place of an "uninterrupted" model, he proposed the idea of "epistemological ruptures".[2] "The history of science", he claimed, consistently displays breaks or leaps in its "line of development", and changes in the idea of reason and rationality occur as a result of these epistemological ruptures.[3] Bachelard's new "scientific spirit" implied that new scientific ideas might contradict "common sense", and that breaks or leaps in scientific thinking tend to go hand in hand with scientific creativity. These leaps are often blocked by epistemological obstacles; these obstacles represent inertia.

The implications of Bachelard's ideas for the development of science were revolutionary, since he was the first to see *episteme* in an historical-developmental context. But his ideas were also revolutionary from another point of view, since he coined the term "psychoanalysis of science". Psychoanalysis of science acquired in Bachelard's thinking two different meanings: On the one hand, Bachelard conceived of an indirect and secondary psychoanalysis that "seeks for the unconscious under the conscious, the subjective value under the objective, the reverie beneath the experiment". "One can study", says Bachelard, "only what one has first dreamt about".[4] This kind of psychoanalysis deals with

< WHY A PSYCHOANALYSIS OF SCIENCE? >

creative, pre-scientific ideas before they acquire their final, cognitive shape; in this analysis Bachelard uses Freudian as well as Jungian concepts. But, on the other hand, Bachelard looked upon psycho-analysis in a different, more cognitive way as well. When referring to epistemological obstacles, he pointed out concepts and methods that involve tacit, unspoken assumptions. By bringing these hidden ideas out into the open, he says, we operate as psychoanalysts of science.

Bachelard's concepts of epistemological rupture, blockage and breakthrough were obviously taken from psychoanalytic theory. The unconscious operates in quanta, in ruptures. The idea of epistemolog-ical blockage is equivalent to the idea of resistance to the uncovering of repressed material, and the aim of the psychoanalyst of science to uncover repressed scientific ideas parallels the uncovering of repressed clinical material of the analysand by the analyst. One critical difference between Bachelard's approach and that of, say, Freud is that Bachelard referred to a cognitive unconscious and not to the unconscious of drives and affect that Freud described. Thus, Bachelard distinguished between a psychoanalysis of creative processes, which handle the emotional unconscious, and a psychoanalysis that uncovers the unconscious intel-lectual products of scientific work in its final shape.

Bachelard's cognitive unconscious has much in common with the unconscious as it was understood and formulated later by Claude Lévi-Strauss and by Michel Foucault. According to Lévi-Strauss, the human mind tends to transform natural reality into an unpredeterminated system, and facts are therefore of a "logical" not natural order.[5] In other words, due to the mediation of human activity and cultural institutions, facts become signs.[6] Thus man communicates via symbols and signs, which for the most part operate unconsciously. Lévi-Strauss adopted the Freudian idea that the unconscious expresses more significant things than the conscious. For instance, according to Lévi-Strauss the cultural unconscious refers to logical structures, while according to the Freudian notion there is no basic distinction between the hidden motives of the individual mind and that of our collective and social insti-tutions. Thus, had a Freudian psychoanalysis of science been developed, it would not have distinguished between the individual scientist's emotional unconscious and the unconscious of the collective. Both, as Freud saw it, are grounded in the same basic motives.

So the differences between Lévi-Strauss's and Freud's concepts of the unconscious are greater than their similarities. For Lévi-Strauss the unconscious represents form, not emotional content. Its function is to impose structural laws upon psychic content, which is content that orig-inates elsewhere and is, on its own, unarticulated.[7] The notion of the unconscious as a structuring activity, and the related emphasis on form

< LOCATING THE IRRATIONAL, UNDERSTANDING THE REPRESSED >

over content, brings into focus the fundamental differences between structuralism and psychoanalysis.[8]

Returning to Bachelard, one can say that he oscillates between a Freudian and a structuralist concept of the unconscious. He applies his concept of the unconscious in the Freudian and Jungian sense at the creative, so-called pre-scientific stage, and applies it in the structuralist sense when referring to the scientific product.

Moving past Lévi-Strauss's structuralism, we arrive at Michel Foucault, who adopted the structuralist notion of the cognitive unconscious, and who, indirectly, laid claim to a psychoanalysis of science. His notion of an archaeological uncovering of the deep structures of knowledge resembles Bachelard's idea of a psychoanalysis of knowledge. He focused, he said, on the unconscious of knowledge, which has its own rules just as the unconscious of the human individual has its rules.[9]

Foucault was interested in the unseen, hidden, obscure and neglected in history. He thought beyond the manifest layers of history. In his introduction to *The Order of Things*, he wrote:

> I did not operate at the level that is usually that of the historian of science . . . the history of science . . . describes the processes and products of the scientific consciousness. But, on the other hand, it tries to restore what eluded that consciousness: the influences that affected it, the implicit philosophies that were subject to it, the unformulated thematic, the unseen obstacles. It describes the unconscious of science. This unconscious is always the negative side of science — that which resists, deflects or disturbs it. What I would like to do, however, is to reveal a positive unconscious of knowledge: a level that eludes the consciousness of the scientist and yet is part of scientific discourse.[10]

Here Foucault in fact defines the archaeology of knowledge. He is writing about history as a discourse, by which he means a set of linguistic practices that generate social and cultural activity and are governed by unformulated (unconscious) rules. Foucault's interpretation of history, like Bachelard's, is characterized by discontinuity and obscurity. While the archaeology of knowledge uncovers the discursive rules, genealogy explains the history of discourse.

In summary, we can say that, whatever their differences, Bachelard, Lévi-Strauss and Foucault approach the analysis of culture through an analysis of a cognitive unconscious.

Before going any further, it should be mentioned that, when juxtaposing the cultural cognitive unconscious with the emotional unconscious of Freudian psychoanalysis, we have to keep in mind that the dichotomy between them is quite artificial—on both a personal and collective level. Both kinds of unconscious have in common the notion

< WHY A PSYCHOANALYSIS OF SCIENCE? >

of the hidden, internal dynamics of things, the unseen, which neverthe-less plays a major role in our psychic and cultural life. We can assume that cognitive elements of the unconscious always have their emotional counterpart; that epistemological ruptures cannot be explained solely on intellectual grounds. If, as Foucault showed, mankind did not examine itself scientifically before the eighteenth century, the reason for this lack of self-critical awareness goes beyond an analysis of the cogni-tive element involved. Self-reflexibility is always deeply rooted in emotional life, and humankind's lengthy blockage in this regard requires an explanation that accounts for cognitive and emotional histo-ries at once. Our notion of the psychoanalysis of science therefore has to be revised in order to include drives and emotions. In this book I intro-duce a psychoanalysis that takes into consideration the unconscious drives and affects introduced by Freudian psychoanalysis without over-looking the cognitive elements of the unconscious that were dealt with by Bachelard, Lévi-Strauss and Foucault.

Apart from Bachelard and Foucault, there are no comprehensive approaches to the deeper layers of the history of science. Nevertheless, there are approaches to history as a discipline that can be accurately described as psychoanalytic and introspective in orientation. Referring to the history of psychoanalysis, Jan Ehrenwald noted:

> The historian of psychoanalysis cannot be satisfied with tracing its origin in the written records of human endeavour, thought and action. Nor should he confine his efforts to the vast system of explicit verbal statements and formulations, which make up the analytic systems of thought . . . he must probe beneath the hidden content as it were. He must look for the hidden motivations, for the forgotten sources and anonymous influences . . . he must be alert to the pres-sure of the unconscious determinants and daily residues derived from proceeding centuries—if not millennia—he may even come across evidence of scotomatisation and reaction-formation which one would rarely expect to encounter in the lofty sphere of scientific theory.[11]

Ehrenwald's statement constitutes a rare declaration of principles within the history of science, one that might easily be adopted as a set of guidelines, not just for the history of psychoanalysis, but for the history of science on the whole.

The first introspective approach to history that I wish to discuss is mentality history. In *Ideologies and Mentalities*, Michel Vovell says that the mentality-based approach developed from a kind of history that moves beyond the individual and relates to culture—behaviour, gestures and attitudes—and to the unconscious, collective expressions of humanity.[12] The 'history of mentality', as he and others define it, is involved with motherhood, love, sexuality and death in history.[13] It includes unformulated mental realities, unconscious motivations and

< LOCATING THE IRRATIONAL, UNDERSTANDING THE REPRESSED >

realities that might at first seem to be meaningless. It leads back to memory and forms of resistance—to the force of mental structure. Vovell defines the history of mentality as the study of mediation between the objective conditions of human life and the ways in which people narrate and live it.[14]

Two basic assumptions in the history of mentality are that a collective unconscious and collective imagination are significant realities to be dealt with in the study of history. These assumptions are not based on psychoanalysis or on Lévi-Strauss's anthropology. They are explained by Phillipe Ariès as empirical ideas referring back to the autonomy of a collective mentality experience.[15] (For all that, however, it must be said that the definition and description of the collective unconscious and imagination offered by Vovell and Ariès are not sharply defined, but remain somewhat obscure. It is not clear, for example, how they differ from other definitions of the unconscious.)

Vovell further explains that the history of mentality turns the modern historian into an anthropologist of the past whose aim is to reconstruct the history of myth and the working of the collective imagination. The history of mentality is a history of unspoken subconscious or unconscious things. Death plays a major role in this exploration by demonstrating silent subjects.[16] The passage from one structure to another takes place in a wide historical framework (*longue durée*), wherein imperceptible changes are of more importance than visible processes.[17] The process of evolution in the *longue durée* is, says Vovell, unconscious, that is, it is not perceived by those who experience it.[18] From Vovell's descriptions and definitions it can be deduced that the history of the *longue durée* has some similarities with Freudian analysis, mainly because it occupies itself with unseen and unconscious processes from the emotional point of view.

A different sort of attempt at breaking with the conventions of historical research in order to analyse history in depth is taken by the analysts of rhetoric. Dominik LaCapra is in favour of leaving behind the fetish of archival research in order to discover neglected facts and phenomena. He prefers an interactive model of discourse that would facilitate an interchange of 'documentary' and rhetorical dimensions of language and, in the process, turn the study of history from a science into a hybrid kind of inquiry, a combination of science and art. Historians, says LaCapra, are fearful of paying attention to their own rhetoric and to the role of rhetoric in science; prefer to see their science as purely logical in its construction. LaCapra's work here refers to the hard sciences as well.[19]

LaCapra explains the difficulty involved in a reconceptualization of historical self-reflection by saying that the relationship of the historian

< WHY A PSYCHOANALYSIS OF SCIENCE? >

with history is a transference relationship. Transference is basically a process that combines repetition and change, but the historian tends to concentrate on the aspect of repetition, idealizing the ahistorical level of the process. In fact, the analysis of history should, by nature, be unidealized and dynamic.[20] LaCapra means by transference the modified psychoanalytic sense of a repetition—displacement of the past into the present. Transference causes fears of possession by the past and loss of control over it and oneself.

The fixation of the historian on ideologically suspect procedures is related by LaCapra to narcissistic mechanisms reflected in the realization of the past. This fixation involves a wish for a totality that leaves no room for the operation of a mechanism of "working through".[21] The opposition of 'scientific history' to the stereotypical 'other'(myth, ritual, memory) serves as a mechanism of defense and denial which makes it impossible to properly evaluate historical material.[22] In LaCapra's eyes, positivism constitutes a denial of transference.[23]

In the spirit of LaCapra's approach, Hayden White rejects the strict dichotomy of 'scientific' and narrative aspects in the interpretation of history. Every representation of historical phenomena, he suggests, involves reality. The linguistic aspect of inquiry and representation constitutes an inseparable part of historiography: "Stories, like factual statements, are linguistic entities and belong to the order of discourse. Narrative accounts do not consist only of factual statements and arguments, but consist as well of poetic and rhetorical elements by which what would otherwise be a list of facts is transformed into a story. Thus one narrative account may represent a set of events as having the form and meaning of an epic or tragic story, and another may represent the same set of events —without doing violence to the factual record—as a farce."[24] White's position corresponds with Roy Schafer's psychoanalytic vision of reality, which is one that contradicts positivism. Schafer categorizes four different views of the human situation: the comic, the tragic, the ironic and the romantic. All four of them refer to certain general ways of comprehending the form and content of human situations and the changes they undergo.[25] But whereas in Hayden White's post-modernist approach no single interpretation is preferred to another,[26] Schafer maintains that not all possible interpretations are necessarily valid.[27]

Hayden White's relativism does not question the reality of historical events so much as remind us that chronicle alone "can't do justice to those events or that history; we need interpretation as well". Narrative, according to White, is a meta-code, a human universal on the basis of which transcultural messages about the nature of shared reality can be transmitted.[28]

< LOCATING THE IRRATIONAL, UNDERSTANDING THE REPRESSED >

What, then, is the function of the psychohistorian? Henry Lawton distinguishes between the historian and the psychohistorian. Both, he says, depict facts and offer interpretations. The ordinary historian, however, is more concerned with representing reality, while the psychohistorian is more concerned with interpretation of that reality, with the explanation of motive and meaning. To this end, the psychohistorian will tend to emphasize emotion and fantasy.[29] Countertransference, accordingly, is an essential tool in psychohistorical research.[30]

According to psychohistorian Peter Löewenberg, there is a basic intellectual—emotional bond between the historian and the historical material. The historian therefore requires the help of an analyst, who can bring a free-floating attention to the voice of history. In order to find the unconscious connections[31] in history, an impartial kind of attention is needed.

Löewenberg's definition and description of psychohistory seems to me especially helpful, since it involves the unconscious. Historians, of course, often offer interpretations of their material, but these interpretations are not necessarily built on unconscious material. One necessary, but not sufficient condition for an effective psychoanalysis of science, then, is the understanding of the transference (or countertransference) involved in the historian's relationship to his or her material.

I have gone into the above discussions of psychohistory because this is the only psychoanalysis of science that we currently have. In the following chapters I shall propose guidelines for a psychoanalysis of science. At this point, having summarized the state of research in this field, it remains for me to explain how this book will attempt to take the psychoanalysis of science further.[31]

The psychoanalysis of science is a method of exploring scientific development by focusing on irrational elements in the developmental process. These irrational elements in science may stem from conscious as well as from unconscious sources. The irrational elements of conscious origin are mainly of interest to the sociologist and the historian, who would ask themselves, for instance, why a seemingly relevant and important scientific idea was at a given time ignored by the scientific community. This kind of refusal to 'see' involves a conscious decision, even though it may be irrational in origin. The analyst, on the other hand, would be mainly interested in blind spots, that is, situations in which it becomes obvious that an occurrence slipped by the awareness of the scientific community. It can be asked, for instance, why totemism held on till the twentieth century and how it affected animal behaviour research.

Even though psychoanalysis is specifically interested in the uncon-

< WHY A PSYCHOANALYSIS OF SCIENCE? >

scious irrational elements in science, the psychoanalysis of science could and should function within an interdisciplinary context, namely, the sociology, history and philosophy of science. Each of these fields concerns itself with the development of discoveries, findings and ideas. Each studies the context in which these ideas develop; it examines the mutual influences within and between particular scientific disciplines and the social environment. Each asks what causes theories to rise and take shape, what obstructs their development, and what causes them to fade into obscurity.

In analysing obstructions in science, we are especially interested in developmental disturbances, about which we may talk in terms of scientific symptoms and scientific stagnation. Thomas Kuhn alluded to their existence more than thirty years ago. He wrote about the proliferation of many versions of one theory as a sign of the failure of normative sciences to solve riddles. Too many versions of one theory are symptomatic of crisis. In this spirit we may claim that too many versions of one theory are a block on creativity. Kuhn also mentioned the distress of perception, meaning the need to deny what one observes. This, he explained, was an obstacle to bridging the gap between what one perceives and thinks and what is permissible in the scientific community. Kuhn demonstrated that new paradigms are usually preceded by a period of professional insecurity stemming from the failure of normative science to solve riddles.[32]

We may continue Kuhn's line of thinking by saying that we discover scientific diseases in the wake of scientific symptoms, such as delay or arrest in certain branches of science and lacunae, or blind spots (to which we shall refer in detail later on). While Kuhn stressed the psychopathology of science in the context of paradigmatic crisis, it is my contention that one cannot properly consider the psychopathology of science without taking into account the irrational elements that often underlie scientific thinking.

Behind so-called objective, scientific thinking, for instance, we may detect the existence of myths and prejudices, the influence of *Weltanschauung*, blind spots and influences from our daily life and the everyday life of science. We may talk, therefore, in terms of a psychopathology of scientific everyday life. The irrational aspect of science is an integral part of our creative scientific self, and at the same time it is also the cause of the defensive aspect of scientific thinking. We cannot split off our scientific self from our personal self. Without recognizing the part played by the irrational and particularly the unconscious in science, we are doomed to fail in detecting many of the obstructions in scientific development.

Obviously, a psychoanalytic exploration of science includes not only

< LOCATING THE IRRATIONAL, UNDERSTANDING THE REPRESSED >

recognizing obstruction (pathology) but also examining creative aspects of science. In this latter category we might discuss the nature of personal knowledge —including character and *Weltanschauung*—and its influence on science, or the influence of culture on scientific research. But before proceeding any further, however, we need to examine the notion I have put forth that irrationality offers an entry into a fresh understanding of scientific development, for the notion is essentially alien to the scientific community.

The psychoanalysis of science would have little if any rationale if we assumed that science is completely objective, that science, scientific truth, scientific observations and conclusions are wholly free of subjective contamination, whether by individual scientists, the scientific community or the society in which science develops. Objective scientific discourse would imply the maintenance of impersonal scientific arguments that could be judged from an external standpoint and be free from personal or collective prejudices. The arguments of different philosophers of science in favour of a subjectivistic approach, however, may build the foundations for a psychoanalysis of science.

The discussion on the place of irrationality in science has been conducted on different levels and from different points of view. Most basically, philosophers have asked the following: What are the facts that we observe and note? Can it be that we falsify them unconsciously? Do we use them selectively within the framework of a given theory? In view of questions such as these, some philosophers of science deny the existence of objective facts, for they see them as always laden, or influenced by, subjective human motives.

Through the 1950s, the history of science was essentially an 'internal' history of science. This history assumed that science developed independently, as though through an inner life of its own. The social and cultural conditions surrounding the scientific community did not, it was thought, influence scientific content, structure, development or theory. One of the few exceptions to this 'internal' historical approach was proposed by Ludwig Fleck, who, by the 1930s, had come up with the idea that the scientific content is conditioned by psychology and by the history and sociology of ideas.[33] Fleck did not agree with the prevailing view that the sole or even most important task of epistemology is in examining concepts for their consistency and interconnections within a system. He believed that the time had come to speak of a comparative epistemology, since every system of knowledge existed, and should therefore be treated, only within an historical context.[34]

In his *Genesis and Development of a Scientific Fact*, Fleck argues that once a closed system of opinions has been formed, it offers enduring resistance to anything that contradicts it.[35] In the history of science, no

< WHY A PSYCHOANALYSIS OF SCIENCE? >

formal logical relation exists between conceptions and evidence. Concepts are not logical systems but stylized units, that either develop or atrophy. Every age has its own dominant concepts. Fleck also coined the term "thought collective", which he applied to scientific processes:

> What is already known influences the particular method of cognition; and cognition in turn enlarges, renews, and gives fresh meaning to what is already known. Cognition is therefore not an individual process of any theoretical particular consciousness. Rather it is the result of a social activity, since the existing stock of knowledge exceeds the range available to any one individual.[36]

Though he was clearly one of the major pioneers in the field, Fleck remained an obscure figure outside mainstream scientific discourse till the appearance some twenty or thirty years later of Norwood Hanson, Michael Polanyi and especially Thomas Kuhn, whose work evoked a storm of scientific dispute that has not yet subsided. Apparently the time was not ripe for Fleck's innovative idea. (It would be an interesting project to explore just why it was not ripe, which is a question that might have implications for the psychoanalysis as well.)

Both Norwood Hanson and Paul Feyerabend are convinced that what scientists call facts are essentially ideational observations shaped by prior knowledge. According to Feyerabend, "Science knows no 'bare facts', as facts are already viewed in a certain way and are therefore essentially ideational."[37] Hanson talks about the paradox of common observation resulting from the isolation of scientists within an observed world, in a way that is consonant with the accepted theoretical beliefs. According to him, theoretical differences in a given domain are not attributed simply to differing interpretations of the same observational data.[38] Michael Polanyi claimed that facts cannot be distinguished from theories, and that all knowledge is in some way personal (subjective): "Some trace of hidden personal bias may systematically affect the results of a series of readings when relating to the selection of facts." He introduces the notion of scientific passion: "As only a tiny fraction of all knowable facts are of interest to scientists, scientific passion serves as a guide in the assessment of what is of higher and what of lesser interest."[39]

The next point in our discussion of irrationality in science is scientific creativity. There is a consensus between objectivists and subjectivists that creativity in itself is personal, subjective in nature. Thus, Peter Medawar describes intuition in scientific thought as non-logical (outside logic). He talks about the sudden appearance of new thoughts and the wholeness of the conception.[40] And Karl Popper admits the existence of subjective knowledge, adding that there is even a need for a theory of such knowledge. Still, in his eyes such a theory would be part

< LOCATING THE IRRATIONAL, UNDERSTANDING THE REPRESSED >

of empirical science, not part of the logic of science or epistemology, since its theme is the growth of someone's knowledge.[41]

The crucial difference between objectivists and subjectivists lies, in the first place, in the distinction between personal knowledge and public scientific knowledge, and, secondly, in the sharp distinction between the creative phase and the phase of scientific proof. Thus, both Medawar and Popper plead for a clear distinction between discovery and justification as two separate and dissolvable episodes of thought.[42]

Feyerabend, on the other hand, rejects the distinction between the context of discovery and that of justification and, accordingly, between epistemology and psychology. According to him, if there are no bare facts, the dividing line between the creative and the methodological phases is artificial. In refuting the notion that we first have an idea or a problem and then we act, he draws an analogy from the world of children, for whom he claims that playful activity is an essential prerequisite to the final act of understanding. He finds no reason why this mechanism should not be applied to the world of adults.[43]

Nor do Popper and Feyerabend agree on the relation between the subjective knowledge of the scientist and objective knowledge in science. Popper claims that, while scientists may be subjective, science is objective. According to him, there is truth in our state of mind, knowledge or belief, and there is objective truth that corresponds with facts. Knowledge in the objective sense, he says, is totally independent of any individual's claim to (objective) knowledge.[44] This is knowledge without a knower, without a knowing subject. But to Feyerabend, for whom there is no universal reason, unreason cannot be discounted: "The debate between science and myth has ceased without having been won by either side for science is much closer to myth than a scientific philosophy is prepared to admit."[45]

When we approach the validity of theories, we are again faced with a divergence of opinion on the question of whether there are absolute criteria or whether the issue approaches the sphere of sociology. In his theory of refutation in science, Popper says that we learn from our mistakes. By finding out that our conjecture was false, we learn more about the truth and get nearer to it. Thus our theories change every time we get closer to our goal. Truth is our guidance.[46] For Kuhn, on the other hand, validating scientific theories is not a question of epistemology but one of sociology. In his view the scientific community preserves its established science in the same way that a family guards its traditions. Thus, he says, scientific truth has more to do with social consciousness or social unity than with objective truth. Scientific method is conservative and subjective, and scientific revolutions try to save it (temporarily)

< Why a Psychoanalysis of Science? >

from stagnation. Changes come only when stagnation leads to too much uncertainty.[47]

Historical studies, too, have demonstrated that scientific theories have been conceptually influenced in form and content and by the cultural milieu in which they have been developed. Amos Funkenstein, for example, has demonstrated this in the relationship between scientific creativity and theology in the Middle Ages.[48] Gerald Holton has shown how Copernican physics was influenced by the idea that the universe reflects the perfection of God.[49] Both he and Lewis Feuer showed how Niels Bohr's idea of complementarity was influenced by William James's work on consciousness and Søren Kierkegaard's concept of dialectical reasoning.[50]

As a result of the blurring of the internalist/externalist dichotomy, histories of science have tended to focus on the creative activity of the scientific community rather than on that of the individual scientist. Meanwhile the debate around epistemology went on.

In the eighties and nineties, objectivist epistemologists abandoned the romantic orientation of truth and began moving in the direction of a more naturalistic orientation. Practical, not moral, questions were emphasized.

Today, both in the natural sciences and in the hermeneutic disciplines, scientists tend to ignore the traditional barrier between the sociology of knowledge and epistemology proper. Thus, objectivists like William Barthley speak in terms of 'moderate truth', in which the goal is not to produce hermetic theories, but to reduce the instance of error. Should we consider such moderate rationalism a quasi-recognition of the irrational elements in science? L. Laudan replaced the Platonic criteria of rationality with a criterion of rational problem-solving. The scientist, Laudan says, should evaluate the adequacy of a theory on the basis of its ability to solve conceptual and practical problems.[51]

Joseph Rouse tells us that the literal sense of the term "discipline must be taken seriously to appreciate that it is a form of political as well as epistemic order". Knowledge, he suggests, has an economy as well as a social and political structure.[52] And Barry Barnes, in his book on Kuhn and social science, noted that most epistemologists aspire to be moralists: They moralize with, and about, the term science.[53]

Taking a slightly different track, the sociologist Steven Shapin describes scientific communication in terms of trust. Only a small fraction of a scientific person's knowledge is acquired by personal experience, Shapin points out. In order to carry out scientific work, they must rely on the knowledge supplied them by the scientific community. And this reliance implies trust. Trust is a necessary component in any

< LOCATING THE IRRATIONAL, UNDERSTANDING THE REPRESSED >

body of knowledge. Social knowledge and natural knowledge are inseparable, and the 'truth' is therefore of intersubjective meaning.[54] In this connection Anthony Giddens writes about the shift in the modern world from face-to-face interaction to trust in institutions.[55] And Mary Douglas goes as far as saying that the construction and maintenance of a system of knowledge can be treated in the same way as any other collective good.[56]

The foregoing account of the Kuhnian/Popperian debate is not in any way meant to represent a full picture of the current ideas on the subject, but to demonstrate that externalists have gained increasing support from the meta-sciences for their arguments. These arguments may serve us as a foundation for the hypotheses on which a fully developed psychoanalysis of science might rely.

Within the framework of a psychoanalysis of science, one's basic explanations can be arrived at by treating scientific texts hermeneutically, that is, by reading between their lines. Beneath the manifest scientific text, we may expect to find the irrational elements that are available to psychoanalytic interpretation.

My assumption is that if one takes a bird's-eye view of the scientific texts, one will sooner or later come across narrative aspects that betray subjective motives. Four elements can be used as guidelines: metaphors, scientific parapraxes, lacunae and myths (mythologies). Together, these four elements constitute the narrative part of the scientific text, in contradistinction to the functional part, which communicates the 'pure' scientific message. Metaphors belong to the rhetoric of the scientific text. They give us insight into scientists' way of thinking, their personal style, their vision and knowledge of their science and the way these sometimes conflict.

Scientific parapraxes include all kinds of contradictions, paradoxes, truisms, *petitio principii* (begging the argument), repetitions and other irregularities in the scientific text. The idea is not to study them from the logical point of view, but to uncover possible unconscious reasons for their appearance.

Lacunae are missing areas in scientific research. The lacunae which are of particular interest for psychoanalytic research are unconscious lacunae. This topic will be dealt with at length in Chapter 2.

Myths and mythologies are an integral part of science. Scientific myths can be detected and interpreted in order to decipher unconscious codes in scientific texts.

According to classical definition, myths are pre-scientific attempts to interpret phenomena, whether real or imaginary. The term mythology metamorphosed from 'real story' to 'invented story'. When we refer to myth in science, we do not necessarily mean that it is related to a story,

< WHY A PSYCHOANALYSIS OF SCIENCE? >

but that our guidelines are not purely rational so much as based on our images, which are often quasi-axiomatic. (According to Aristotle, axioms stem from the primitive.) Scientific texts may contain myth in the sense of story (narrative), and myth in the sense of quasi-axiomatic images.

Myths and mythologies in science are sometimes approached as irrelevant or false aspects of science. When this is the case, we sometimes wish them away or ask ourselves why they have been brought in. Again, sometimes the myths and mythologies in science can be related to as creative elements. Either may be the case depending on the context. If we assume that irrationality is an integral part of ourselves—including our scientific work—we must deal with myth in science as another scientific fact, taking it into account and finding out how it influences science for good and for bad. In this context, attempting to overlook or isolate myth would not make our science look more objective. Indeed, the capacity to recognize irrational elements and discover how they influence scientific thinking may broaden the scope of science.

The place of myth and mythology in human culture has engendered a 200-year-long intellectual argument between two philosophical schools. The Enlightenment granted logos a central place in cultural life and set out to destroy myth as harmful and irrelevant to advanced civilization. Romantic philosophy, on the other hand, sought to return myth to a respected place next to logos. The debate on the place of myth in culture and in science is still going on, and will no doubt continue in the twenty-first century.

In philosophical terms, one may locate the clinical aspect of Freudian psychoanalysis in the Romantic school, and its structure and anthropological philosophy in the Enlightenment. This does not mean that Freud set out to actualize the vision of Romantic philosophy in science. On the contrary, this occurred as an unwanted product when psychoanalysts found that the mythological layers comprise an integral part of both individual and societal culture. To this day psychoanalysis unknowingly continues the search after the 'real self' that was begun by Vico, Herder and Schelling.

Giambatistta Vico, who preceded the Enlightenment and Romantic movements, tells us that "myths are not false narratives, nor are they allegories. They express the collective mentality of a given age."[57]

Romantics believe that myth is necessarily imposed on us from within by the very nature of consciousness. This consciousness is the true active subject. According to Schelling, 'Myth is a special sort of reason reducible to nothing else.'[58] This ideology is the forerunner of the primary processes as revealed by Freud, which bear a mythical character in that they do not answer the requirements of ratio but are attuned

< LOCATING THE IRRATIONAL, UNDERSTANDING THE REPRESSED >

to the subject itself. The fact that the primary processes are dictated is interpreted in psychoanalysis as the determinism of the unconscious. In other words, the primary processes are viewed in psychoanalysis as the real subjects, and as bound to the authentic core of human beings.

Despite Freud's anthropology and his anticipation of the ruling of ratio, his writings indicate that he did not entirely devalue myth. For example, in a letter to Albert Einstein,[59] he admits that instincts are a kind of mythology, contradicting his view of the absolute truthfulness of logos in science. And in a letter to Wilhelm Fliess, he describes culture as a projection of our inner life, implying that mythology is the cradle of culture.[60] Still, according to Freudian anthropology, the more a culture is developed, the less it needs mythology.

We find a tremendous shift in the equilibrium of myth/logos in Freudian psychoanalysis. It seems that the same school that seduced us into believing in the centrality of myth to the development of culture and the individual also contends that, in a completely developed culture, myth would have no place in the psyche of the individual or the community. In according myth seemingly substantial importance and then endeavouring to push it into the abyss, it sometimes seems that Freudian psychoanalysis seeks to suppress the fact that myth represents both the creative process and the occult—in other words, it seeks to deny the complementarity of myth and logos.

According to the philosopher Ernst Cassirer, classical psychoanalysis turned myth from a living element of culture into a kind of preparation for intellectual observation, and by doing so made it 'perfectly logical, almost too logical'.[61] Thus, while thinkers in other fields—philosophers like Cassirer,[62] historians like Hans Blumenberg[63] and anthropologists like Claude Lévi-Strauss[64]— were able to find a new modus in the equilibrium myth/logos, Freud could not do so. The result is that Freud made exclusive use of mythology as a cornerstone of psychoanalytic theory, yet devoted scant space to myth and mythology in enlightened cultural life.

In my view, although some of science's basic assumptions are in fact myths, this does not mean that our scientific myths *a priori* have no basis. Rather, it means that their method of persuasion is not based on logical-scientific systems of verification.

The myth of scientific truth, for example, says that we can achieve ultimate truth through the scientific method, or that we are on the way to such truth, although we shall never reach it. When Karl Popper says: "We have no criterion of truth but are nevertheless guided by the ideal of truth as a regulative principle, "[65] he is not aware that he is offering no evidence of the creature called 'truth' or of how we can be guided by it. To him it is axiomatic that we can approach truth; however, it appears

< WHY A PSYCHOANALYSIS OF SCIENCE? >

more like a *petitio principii*, a form of secular theology, especially as it is approached but never reached, in the same way that theological texts approach God but are never able to reach him.

The idea of human advancement, to take another example, is a myth, not because it is necessarily untrue, but because it, too, has never been proven. It is a classical idea of the Enlightenment that Ernst Heckel formulated in the postulate "ontogeny recapitulates phylogeny", [66] and that Auguste Comte expressed in his three stages of human development: animism, metaphysics and the acknowledgement of reality.[67] As a latecomer to the Enlightenment, Freud used this concept as a guideline in his anthropological philosophy as if it were axiomatic. It aided him in demonstrating that knowledge and the realization of objective fact are what render individuals enlightened. Comte's idea that mankind developed in stages, advancing from the primitive to the enlightened, had far reaching consequences for the development of psychoanalytic anthropology. It prevented psychoanalysis from observing and evaluating cultures for what they are because it prejudiced analysts with the unproved idea of advancement.

A third myth is that of the idea of scientific unity, or in other words the idea that the unification of many phenomena and postulates into one single theory is logically and aesthetically desirable, and serves the cause of simplicity, elegance and scientific coherence. This is the case so long as such a process of unification is plausible and transparent, yet we have to keep in mind that the idea of scientific unity might serve as a good in itself instead of being a means for scientific clarity. When an attempt to explain all phenomena within one central idea becomes compulsive, we know that we can detect mythical (or religious monotheistic) thought behind it.

Having discussed three guiding elements—metaphors, scientific parapraxes and myths—we arrive now at the fourth element, the lacunae, to which a special chapter will be devoted.

< LOCATING THE IRRATIONAL, UNDERSTANDING THE REPRESSED >

2

Lacunae in the Development of Science

What we want to know from our patient is not only what he knows and conceals from other people; he is to tell us too what he does not know. — S. Freud

Theoretical contradictions, defects, lacunae, may indicate the ideological functioning of science. — M. Foucault

The commitments that govern normal science specify not only what sorts of entities the universe does contain, but also, by implication, those that it does not. — T. Kuhn

This chapter analyses the development of science on the basis of a new hypothesis, namely, that unconscious lacunae play a significant role in the history of science and are crucial to our understanding of scientific development. The history of science usually deals with the facts that constitute the raw material necessary to explain the origin and development of discoveries, findings and ideas. Historians of science study the context in which these facts develop, their mutual influences within and between particular scientific disciplines, and within the social environment. They ask what causes theories to rise and develop, and what causes them to fade into obscurity.

But they never distinguish the lacuna as well as the substrate that constitutes it. Lacunae make up an integral part of the scientific construction, in that they allow us to see the history of science from two points of view: that of what happened in science, and that of what did not happen (questions not asked, blind spots).

There are two kinds of questions that are significant in this respect. The one, and better known, lies within the region of conscious elimination or avoidance of scientific matters due to the *Zeitgeist*, or the preferences, interests and selective curiosity of the individual scientists or the scientific community. The choice begins with paradigms.[1] Like

< LACUNAE IN THE DEVELOPMENT OF SCIENCE >

the choice between competing political institutions, that between competing paradigms proves to be a choice between incompatible modes of community life.

The process of selection in science begins with an idea or question that comes to the mind of the scientific community and continues through to the solution of that idea or question. But what comes to the mind of the scientific community is influenced by the inhibitions and restrictions that society places on it out of political, economic and religious considerations. And, indeed, the scientific community has the power to silence the thoughts and actions of the individual scientist. Factors such as these have always acted to delay inquiry into scientific ideas and projects that are not considered beneficial to society for one reason or another. In addition, the scientific community itself has always, consciously and unconsciously put spokes in the scientific wheel.[2]

The conscious resistance of scientists to certain discoveries sometimes causes the latter to fade into oblivion; scientific development is thereby arrested. Sometimes, however, this resistance is unconscious. Thus, when subjects that should be considered important are not researched, despite their relevance and their lying, so to speak, under the nose of the scientist, it may be asked whether this is due to unconscious repression on either the part of the unconscious of the individual scientist or the scientific community. Yet the question of what leads a scientific community to overlook central and important issues and phenomena has not been touched upon by historians and philosophers of science. Some examples of scientific lacunae follow.

The paucity of scientific experiments and the general lack of interest in technological development in ancient Greece can serve as an example of the lacunae of science. Shmuel Sambursky, who claims that this lack or interest resulted in a lack of scientific creativity, rationalizes the phenomenon by stating that, in those times, the process of disconnecting science from myth had not yet been completed. Because the ancient Greeks viewed the cosmos as a living organism, to which they felt close, they resisted damaging the natural, preferring to observe nature from afar. Therefore, they had no inclination to perform experiments.[3]

Such abstention from certain ways of inquiry had far-reaching consequences. The history of the discovery of black holes can illustrate this point. Stephen Hawking, the theoretical physicist, tells us that the notion of black holes in the cosmos was actually born about two hundred years ago, when John Michell claimed that a sufficiently condensed star would have a gravitational field so strong that it would prevent light from escaping from it. (Many stars of this sort exist, although we cannot see them because their light is reabsorbed by grav-

< LOCATING THE IRRATIONAL, UNDERSTANDING THE REPRESSED >

itational force.) A short time later, Marquis de Laplace came up with a similar idea, which he included in the first edition of his *World System*, but he left it out of later editions. Hawking assumes that he did so because the idea seemed too "crazy" to him.[4] Almost two centuries passed before the idea was brought up again by the astrophysicists. It would be profitable to ask why a seal of silence was imposed on this idea for such a long time.

The resistance of scientists to scientific discovery has been studied by Bernard Barber. Resistance can originate from different sources, he says, but the consequence of it is that the discovery will not become known for a while, or possibly forever, and a lacuna is thereby formed in science. In this context, he mentioned Mendel's discovery of the laws of genetics being put aside for several dozens of years for reasons which could hardly be considered scientific. Barber mentions the suspicion of botanists toward the use of mathematics, Mendel's low scientific status ("the unimportant monk from Bruenn"), and the *idée fixe* about the rules of inheritance. Mendel's conception of separate inheritance of characteristics ran counter to the predominant conception of joint and total inheritance of biological characteristics. It was not until botany changed its conception and concentrated its research on the separate inheritance of unit characteristics that the theory was rediscovered. For forty years these important biological laws were ignored.[5]

Another example of a lacuna in biology is the Soviet denial of the existence of chromosomes because recognition of this biological fact would endanger Marxist philosophy, according to which it is only environment that is responsible for changes in the human being. This is why only an ideal society (communist society) can change the individual in the right direction. Biology in the USSR was ideologically shaped, mainly under the direction of Trofim Denisovich Lysenko.[6]

Medical authorities treated the discovery of Louis Pasteur, the father of microbiology, with contempt, in part because he was not a physician but "merely" a chemist.[7]

In the field of neuropsychology we find a lacuna in the case of George Gilles de la Tourette, who in 1880 discovered a bizarre and multifaceted syndrome that was eventually named after him. The syndrome involved contortions of the facial muscles, compulsive behaviour, unchecked outbursts of laughter and an exaggerated tendency toward clowning and social provocation. In the wake of Gilles de la Tourette's work, several clinical studies were undertaken in the late nineteenth century, but after that the syndrome 'disappeared' and was neither discussed nor written about in the professional contexts until the late 1960s, when suddenly it became apparent that there were numerous cases of this syndrome. Today, there exists in the United

< LACUNAE IN THE DEVELOPMENT OF SCIENCE >

States an organization of 50,000 people suffering from Tourette's Syndrome.

Oliver Sacks has tried to explain the ongoing silence that surrounded the illness and the sudden awakening to its existence in the 1960s. As he sees it, in the nineteenth century it was sufficient to provide a scientific description of an illness or syndrome, whereas in the twentieth century there was a need to explain phenomena in order to be persuaded of their existence. The syndrome in question was extremely difficult to account for because of the variety of symptoms it involved. Interest in the syndrome was revived when it became apparent that there was a chemical explanation for the illness which came in the wake of the use of a new drug that affected the facial contortions. Even then the chemical explanation still did not explain the many other symptoms of the illness. In any event, Oliver Sacks presented this case, along with others from his neurological experiments, in order to show how labyrinthine the ways of science often are, and how misleading it is to view the history of science as a linear development.[8]

The code of law within the judicial system serves as a classic example of how lacunae shape a system. When lacunae are exposed in the courts because the right match between a specific juridical problem and abstract law cannot be found, it is usually due to legislators having overlooked relevant cases when constructing the law. Lacunae in the law are unavoidable because the law cannot be abstract and comprehensive enough to include every possible relevant situation. While some lacunae result from faulty or incomplete thinking on the part of legislators, there are times when they result not from a lack of abstract ability, but from the promulgation of a defective law that results from a conflict between juridical and political interests. If, in fact, it turns out that the legislator did not think the problem through to the end, how can we determine whether this was a legislative lapse, professional deference or a conscious attempt to bypass the problem? Even if we reconstruct each specific case, or the whole process of legislation in depth, in most cases we would probably not arrive at a definitive answer as to whether the lacuna was conscious or unconscious in origin.

As for the field of psychoanalysis, which evinces great interest in the arts as an area of research, one may ask why it has engaged in so little research in the sphere of science (a question that will be discussed in greater detail in the Epilogue). Marshall Bush claimed that scientific interest as well as scientific creativity had received surprisingly little consideration in psychoanalytic and related literature. He summarized some ideas concerning scientific creativity, but they only demonstrate how scant an interest psychoanalysis took in science as a field of research. Bush mentioned missed discoveries in which scientists failed

< LOCATING THE IRRATIONAL, UNDERSTANDING THE REPRESSED >

to appreciate essential factors that could have led to a breakthrough. The deeper reasons for this phenomenon were not explored.[9]

With regard to history as a lacunary structure, Lévi-Strauss has made some interesting comments on the historical facts chosen for study and their selection. According to him, historical facts are no more given than other facts. They are abstractions, and any attempt to write history must of necessity be selective and leave a part of the picture outside its framework. He wrote, "Every history that claims to be universal is nothing but a joining together of several local histories within which the amount of black holes is much higher than the filled in place."[10]

What is of interest to Lévi-Strauss is not only the holes but their underlying meaning. Thus, he wants to know for whom any specific history is written, because this may help us discover what sorts of lacunae to look for—what this 'history for' wants to conceal from us, either consciously or unconsciously.

Another example of a scientific lacuna that is attributable to developmental retardation is the relatively late development of ethology as a branch of biology. Since all that is needed to observe the behaviour of animals are a pair of eyes, binoculars (invented three hundred years ago) and much patience, it is strange that ethology did not begin *de facto* until Konrad Lorenz, that is, in the middle of the twentieth century.[11]

What prevented biologists from research into animal behaviour? Not lack of interest, because we know that human beings in almost all cultures have always been deeply interested in animals. This is evinced by, among other things, the deep attraction of children to animals, as well as to the manifold tales about animals, in all cultures and many religions. Why, then, were biologists and psychologists so dilatory in studying how animals behave? Recent findings in ethology reveal that what we thought about the behaviour of animals has been semi-mythological, and we may therefore assume that the lacuna involved in the late development of ethology is attributable to an unwillingness to give up our mythological attachment in exchange for the cold light of science.[12]

As to whether lacunae are conscious or unconscious, the difference seems to be clear-cut in some cases and opaque in others. For example, the lacunae in the field of genetics belong categorically to the domain of the conscious, for, in the cases of both Mendel's law and chromosome theory, there was a conscious attempt on the part of the scientific community to prevent these theories from being recognized. In the case of Mendel, this was due to uneasiness in confronting an unconventional idea that might shatter the security of conservative thinkers; the Soviet refusal to countenance chromosomes was political in origin. On the other hand, the neglect of science as a field of psychoanalytic research

< LACUNAE IN THE DEVELOPMENT OF SCIENCE >

may belong to the category of unconscious lacunae because it was not so much planned as 'forgotten'.

In those instances in which it is difficult to categorize a lacuna as unequivocally conscious or unconscious, it may be assumed that it is due to a mixture of both factors. We may therefore never know why Laplace introduced his hypothesis about black holes in the first edition of his book and then dropped it, for he never revealed his reason for so doing. Even if he had, how are we to determine whether the explanation was the real reason? If he told us that, upon further reflection, the idea of black holes seemed to him too crazy to pursue, why is it that this escaped him when he published the first edition? In other words, what were the deep reasons behind his first daring proposal and then his withdrawing of it? Likewise, why did the theory of black holes vanish after appearing over two hundred years ago?

Even when there are obvious conscious reasons for lacunae, this does not mean that unconscious motives are not also involved. When the Pythagoreans concealed the existence of irrational numbers because it shattered their belief in harmony, were they aware of why they were afraid of irrational numbers?[13] It is therefore important to keep in mind that too sharp a division between conscious and unconscious lacunae may be artificial.

How do lacunae come about? We've seen that conscious lacunae involve intentional acts within the scientific community, but are the lacunae due to collective motives, or do they belong exclusively to the category of the individual? This question is of far-reaching significance, for if scientific repression is an individual act, the scientific community should be in a position to correct the scotoma, to realize what the individual has missed or overlooked or repressed, and we could justifiably claim that the scientific community affects the path of science within the restrictions of the *Zeitgeist*. However, if it becomes clear that the scientific community cannot correct individual blind spots, that it either succumbs to them or produces them as a collective, then we would have proof that there is a subjective factor in scientific development.

In suggesting that the scientific community itself produces lacunae, we do not imply the existence of a mystical entity affecting the development of science. On the contrary, we are saying that a certain atmosphere influences the direction of science and guides individual scientists. For despite the influence of the scientific community and society at large, it must never be forgotten that scientific work is an individual pursuit. The distinction between the professional and the private personality is artificial, and it is one of our working hypotheses that the unconscious of the scientist acts to produce lacunae.

In considering whether there may nonetheless be any validity in the

< LOCATING THE IRRATIONAL, UNDERSTANDING THE REPRESSED >

argument that lacunae might be formed by a communal unconscious, we may refer to Karl Popper, who claimed that, while the creative processes of individual scientists may be influenced by their subjective judgement, science itself is objective because objective knowledge is always independent of the subject that creates it.[14] Popper's assumption has never been proven, however, and it comes across as a 'logical leap' that is due to the desire to see science as independent of human beings. In fact, it has the flavour of a secular theology. I intend to demonstrate how the 'truth' Popper speaks of depends on the unconscious. For, in this respect, the scientific community does not differ from the individual scientist. It is not that the scientific community does not always correct the blind spots of individual scientists, but the opposite: individual scientists often see what the scientific community cannot. This is because the rule of the herd applies to the scientific community in the same way that it does in any human community.

When we ask how we know that a lacuna in science is unconscious, we are asking precisely the same question psychoanalysts ask when they come across repressed material in patients: If something is unconscious, how can we possibly know about it? In psychoanalysis we have the general theory of the neuroses, which tells us that the unconscious implies its existence through the appearance of symptoms. In order to prove the existence of unconscious lacunae, we would, then, have to show that some scientific ideas have arisen from the repressed unconscious.

Kuhn's description of science during periods of crisis is given, in psychoanalytic terms, as a disturbance in self-image. He depicts the scientists living through the tension between an existing paradigm and new facts as having the ground cut out from under their feet. In this context, he discusses unconscious scientific resistance to exceptional phenomena and describes the symptoms that can develop as a consequence of the scientific conflict. According to Kuhn, the consequences of the scientific crisis are not completely dependent on consciousness, a comment that reminds us of the fear of change in the psychoanalytic process.[15]

Foucault says that the unseen is reality itself, but it is necessarily unseen.[16] For the repressed energy of voids always evinces the existence of something beneath the surface. And the analytic experience demonstrates that whenever we succeed in overcoming resistance and bringing what is repressed to the surface, what happens as a result differs from the experience connected to the exposition of conscious material. The bringing up of repressed material widens the field of sight. As long as the process of repression continues, 'scientific symptoms' and developmental disturbances remain.

< LACUNAE IN THE DEVELOPMENT OF SCIENCE >

By endeavouring to discover what has not happened in science—by viewing the scientific lacunae as psychoanalytic phenomena—we may better understand the dynamics of scientific development *vis-à-vis* specific situations. For example, when an unconscious lacuna is dissolved, new scientific associations develop in conjunction with the disappearance of inhibitions in the thought process. This freedom allows access to new roads of creative thinking.

However, since lacunae often reveal themselves only by chance, we must be on guard if we are not to let their existence escape us. Even though we are not always in a position to determine their scientific value from the outset, we must have the patience to follow them until we know for certain whether and how they fit into accepted theories. As Kuhn puts it, no theory should clash with one of its individual instances. If we wish a theory to be salvageable, it must be restricted to those phenomena for which we have proof. If we take one more step, we are outside science.

What sometimes looks like a second-rate problem at first glance might turn out to be a significant or even a decisive factor in the development or inhibition of a theory. In this connection, Chaos theory can teach us a lesson. The question on this matter put to the 1979 annual conference of the American Association for the Advancement of Science was: "The Capacity to Prophecy: Could the Rush of the Butterfly's Wings in Brazil Cause a Tornado in Texas?"[17]

< LOCATING THE IRRATIONAL, UNDERSTANDING THE REPRESSED >

3

Psychoanalytic Historiography

The method proposed in these pages assumes that by tracing and analysing scientific texts according to the four categories detailed above (p. 15), we shall uncover repressed scientific material. This is what happens in the psychoanalytic session when blind spots are traced and analysed, and this method may also help us arrive at new scientific ideas. By using psychoanalysis on scientific texts, scientific creativity may be enhanced. The main task I have undertaken in this book is to show that the concept of irrationality in science is not just a deductive philosophical idea, but that it can be tested inductively on scientific texts. If it can be successfully demonstrated that unconscious lacunae can contaminate scientific material, this will show that irrationality does not stem from the subjectivity of the individual scientist alone, but is endemic to science *per se*. Non-hermeneutical sciences are not excluded. The assumption of irrationality in science does not distinguish between different fields of science. After all, the subjective motives uncovered by Kuhn referred to the natural sciences.

Every science is based on texts. While a hermeneutic science obviously provides more metaphors and scientific parapraxes, contradictions and paradoxes can and have been found in natural sciences as well. Both types of sciences are prone to myths and lacunae.

In the chapters that follow I shall demonstrate the described method on psychoanalysis itself: I shall use the method of psychoanalysis combined with history, sociology and philosophy in order to research the science of the psychoanalytic discipline. By uncovering and analysing psychoanalytic texts, I intend to show how this can explain the development and obstruction of certain ideas, and how at the same time new psychoanalytic ideas can be evoked.

Using psychoanalysis in order to analyse psychoanalysis raises a methodological problem. Is this way of doing things scientifically valid? Kurt Goedel's theorem would argue against this supposition. It states that it is impossible to prove a system by using only theorems derivable

< PSYCHOANALYTIC HISTORIOGRAPHY >

from that system. To prove freedom from contradictions it is necessary to use theorems which can only be proved by going outside the system. But in order to prove that the new principles do not conceal contradictions, one must use new principles beyond them. There is no end to the regression. Every language becomes meta-language in the chain of endless proofs.[1]

This endless regression led Bridgman to conclude that we have to resign ourselves to having been born into a world of paradoxes and to live with it. He says that the brain that tries to understand is itself part of a world that is trying to understand.[2] In other words, we are both the observers and the observed. Are we then doomed to irrationalism? William Barthley argues that the inherent irrationality of traditional rationalism stems from its pretence to justifiability and that the problem would be resolved by basing the rational on criticism rather than on justification.[3] I would go even further: it is the rationale recognition of the irrational part of ourselves, including our sciences, that enables us to recognize the methodological aspects of science and thereby demythologize it to some extent. Barthley considers himself a moderate rationalist, which amounts to saying that he is a moderate irrationalist. And, as Joseph Agassi suggests, the more closely one looks at the difference between rationalists (Popper) and irrationalists (Michael Polanyi), the more the difference seems to vanish. The first advocates as much doubt and criticism as possible, yet admits that complete doubt is impossible; the second calls for limiting doubt because doubt is impossible and effective doubt well aimed.[4]

From the above, it becomes clear that I am in good company with my doubts about my methodology and way of working. When psychoanalysis researches itself, it tends to come up with a philosophy which—despite Freud's argument that science is objective—*a priori* finds the irrational in everything. Rational philosophers tend to find proof for the objectivity of science with the same conviction. Beyond my being in good company when I deal with unavoidable paradoxes, my methodology—which I am not the first to suggest—insists that science in general, and psychoanalysis as a private case, should be investigated from an interdisciplinary point of view, that is, psychoanalysis alone cannot effectively research psychoanalysis; it must be used in combination with history, sociology and philosophy.

Psychoanalysis and the Meta-sciences

What is or what should be the place of a psychoanalysis of science within the meta-sciences—philosophy, history and sociology? Should it

< LOCATING THE IRRATIONAL, UNDERSTANDING THE REPRESSED >

be treated as a branch of one of them? As an element of each? Should it be understood as a distinct field of inquiry, with an independent history? An answer to this question would help clarify our approach to analysis and, historically, its oversights.

Two factors lead us to conclude that psychoanalysis of science is not part of philosophy. First, philosophy of science is specifically focused on the validation of the foundations of sciences, on giving science a theoretical framework. It is concerned with the borders of the sciences, scientific credibility and scientific proof. This is *not* what the psychoanalysis of science means to do. The psychoanalysis of science is occupied, instead, with understanding how mechanisms of scientific processes function in depth. Its task is to analyse.

The second reason for the exclusion of the psychoanalysis of science from the discipline of philosophy is that the philosophy operates on the deductive level while psychoanalysis works on a strictly inductive level. Moreover, psychoanalysis operates, at least partially and by definition, on the unconscious plane. Philosophers may, of course, draw conclusions from what a psychoanalyst of science brings to light. And the philosophy of science and psychoanalysis of science share a preoccupation with the relationship between epistemology and psychology.

Clearly one can dichotomize between philosophy and psychology—but only as a philosopher. The psychoanalyst of science must always consider the two fields in relation to one another lest he undermine his essential argument for the validity of his very analysis.

In the relationship between psychoanalysis, on the one hand, and history and sociology, on the other hand, one can find much more common ground. Psychoanalysis is related to the introspective schools of history, such as mentality history and the narrative approach (represented here by LaCapra and White). Neither of these approaches believe in objective history and both of them read history between the lines. Still, they differ from the psychoanalysis of science in two ways: From the structural point of view one may say that psychoanalysis has one foot in history and the other outside it. It belongs to the historical discipline in so far as it analyses the rhetoric of the historical text. It stands outside it when analysing lacunae—an absence. In this respect the psychoanalysis of science can be considered the negative image of history. It deals with what did not happen.

From the conceptual point of view, the psychoanalysis of science stands squarely outside the realm of history. It is closer to the Foucaultian way of thinking, according to which, by thinking one's history, one can free oneself from what one silently thinks and enables it to be thought differently. Expressed in intellectual terms, Foucault speaks of a sort of analytic process of change and working through.

< PSYCHOANALYTIC HISTORIOGRAPHY >

History is not the aim, but the substrata on which this analytic process takes place. This is the common ground between the Foucaultian intellectual analysis and the Freudian psychoanalysis of science proposed here.

Psychoanalytic Historiography

It is pertinent to ask how thoroughly psychoanalytic historiography was treated from the point of view of the psychoanalysis of science. Such a discussion will form a bridge between Part I and the rest of the book, in which psychoanalysis itself will be the subject of research.

The essential question to ask in regard to efforts in the field of psychoanalytic historiography is: Was psychoanalysis analysed? Odd as it may seem for a discipline fundamentally concerned with introspection, psychoanalytic historiography stopped short of analysing psychoanalysis, for psychohistorians analyse history, not psychoanalytic history.

Most of the writings in psychoanalytic historiography are in essence chronological in their nature. A close look at André Haynal and Falzeder Ernst's retrospective of psychoanalysis, *One hundred Years of Psychoanalysis*, which records contributions to the history of psychoanalysis, will give one a good sense of what now preoccupies historians of psychoanalysis. Most of the volume is dedicated to Freud's early work and to his relationship with Sandor Ferenczi.[5] One would imagine that a reliable retrospective would offer some sort of panorama, or cross-section, of the century's scientific occurrences and processes, its intellectual highlights, its influence on other aspects of twentieth-century culture and the way it was influenced by that culture. It would be important to get some notion of how central ideas developed, and how and why they survived or disappeared. But the Haynal–Falzeder volume, like many others, shows how strong the fixation on the founder still is.

In many ways, psychoanalytic historiography reflects itself, in that it tends to concentrate on persons and not on processes. It matters little whether a given historian writes in favour of Freud or against him; the mechanism remains the same—fixation. This fixation takes place not only in scientific biographies, but also in studies of developmental psychoanalysis. As Arnold Davidson put it: "How we characterise Freud's place in the history of psychiatry ought to depend not on who said what first, but on whether the structure of concepts associated with Freud's writings continues, extends, diverges from, or undermines the conceptual space of the 19th century psychiatry.[6] Rather than concen-

< LOCATING THE IRRATIONAL, UNDERSTANDING THE REPRESSED >

trating on 'who said what first', we should be interested in how ideas are transmitted or blocked.

Judith Vita was aware of the fact that psychoanalytic historiography is more a history of scientists than a history of science. She is convinced that in order to make a case for the concept of historical sequence grounded in a demonstrable process of continuous change, one has to reverse the common practice of biographical narration, which tends to display the central historical situation in terms of an individual's development.[7]

Henry Ellenberger's work on the history of the unconscious is a case in point. This historical survey of facts gives the reader an excellent account of the cultural and scientific background of Freud's revelations, and it does so without placing Freud in the centre of the discussion. Still, this work fails to provide in-depth answers to important questions, that is, fails to treat the underlying discourses and motives that enabled such new ideas to grow over the course of the last two centuries. By focusing on the background of Freudian ideas, one very often engages in a kind of closed–circuit analysis. A psychoanalysis of psychoanalysis would address the phenomenon that Eric Erikson identified in his remarking that "we know people whom Freud . . . introduced into psychoanalysis, but psychoanalysis itself [seems to have] sprung from his head like Athena from the head of Zeus". And Frank Sulloway noted that "like psychoanalysis, all sciences hold a theory, however unspoken and implicit, about the proper route of scientific discovery which mythologizes the memory of every great achievement in science". Written psychoanalytic dialogues would be an excellent source for finding out what kind of discourse was taking place, what kind of ideas were included or excluded in that discovery, and on what kind of ideological or psychological grounds.[8]

Finally, before moving on to our own analysis, we shall explore a few examples of how the history of science has moved beyond the biographical and developmental approach. Several writers in the field have adopted what we might call an "archaeological approach", that is, the work is an attempt to uncover 'buried' material. José Brunner's attempt to read Freudian texts between the lines in search for hidden aspects is a good example of this method. In his search for the political Freud, Brunner analyses Freudian texts purely from a rhetorical point of view. He finds that most of Freud's metaphors are essentially political in nature, and concludes that Freud's hidden motives were basically political. Here we have an example of a treatment of hidden aspects, but Brunner's discussion remains bound to a single scientist; it does not extend to his science.[9]

Arnold Davidson demonstrated how one can work with Foucaultian

< PSYCHOANALYTIC HISTORIOGRAPHY >

principles in psychoanalytic historiography. Taking Freud's "Three Essays on Sexuality" as his object of research, Davidson asked whether Freud had come up with a new idea of sexuality in that work. Did Freud's 'conceptual space' continue or break with the conceptual space of nineteenth-century psychiatry? Davidson was looking specifically for the conceptual space of perversion. There was no unanimity about the nature of perversion at that time, but it was generally accepted that perversion was a functional disease of the sexual instincts. It followed that a unified treatment of perversion could be prescribed, since a heterogenous group of phenomena could have been put under a single heading. But by claiming that the human being is basically "polymorphous perverse", that there is, in other words, no natural sex object, Freud undermined the entire structure of the nineteenth-century theory of sexual pathology. From this one could conclude that there is no basic pathology in instincts becoming attached to an inverted object. In short, there is no true perversion. Since Freud destroyed familiar nineteenth-century concepts of sexual instincts and the sexual object, he could not have meant what other nineteenth-century writers meant when he spoke of perversion. Although Freud is using the same terminology as his predecessors (perversion), his thinking here belongs to a totally different scientific discourse. In this new idea of sexuality, Davidson sees the beginning of a change in the culture's collective understanding.[10]

Another interesting example of the Foucaultian approach is associated with the problem of universalization in psychoanalysis. Universalization of the psyche, or psychic unity, as the anthropologists call it, is axiomatic within the psychoanalytic corpus of knowledge. The rules of the unconscious were derived individually via Freud's observations of Viennese patients and were then imposed by him on all mankind. This universalization of Freud's clinical and theoretical conclusions was tacitly accepted by the psychoanalytic community. The notion of psychic unity was never put in question in the psychoanalytic literature. It remained a hidden problem. Although the existence of such a problem was obvious, the problem itself was not articulated. It was, in other words, denied. For a long while historiography did not bring the question out into the open, and so the history of psychoanalysis simply echoed the understanding of psychoanalysis, without seriously questioning the nature of this denial. The history of psychoanalysis failed to articulate the issue.

This important step was recently taken by Sander Gilman in *The Case of Sigmund Freud*. Gilman attempts to explain the phenomenon of universalization in Freud's humanity. He finds that in the medical literature of the *fin de siècle*, Jews were associated with sexual perversion.

< LOCATING THE IRRATIONAL, UNDERSTANDING THE REPRESSED >

According to early twentieth-century medical literature, there was a greater rate of inbreeding in Jewish communities. The high incidence of madness among Jews, as well as their disposition to degenerative diseases, was understood to be a result of their inbreeding. Since incest concerns not only the integrity of the body, but that of the community (*Blutschande*), the Jews were considered a danger to society. Freud, says Gilman, attempted to universalize the argument about the origin of Jewish diseases.

According to Freud, the Oedipus complex, with its murderous and incestuous elements, including its extension from the individual to society, is rooted in the very existence of humanity. As a Jewish physician, Freud was forced to deal with the claim that Jews should not take part in scientific work because of their inherent pathology. By claiming a universal origin for the inherent perversion that the Jews are accused of, Freud was able to counter the anti-Jewish conclusions of *fin-de-siècle* science. In fact, Freud tacitly admitted that the Jews were perverted, but he argued that they were in good company. Compelling as Gilman's arguments are, he examined Freud's writings not for their psychoanalytic content, but for their rhetoric, an analysis of which offers us, as he demonstrated, a message that is strikingly different from the acknowledged "historical understanding" of Freud's thought.[11]

The Sociological Uniqueness of Psychoanalysis

The first uniqueness of psychoanalysis is a veneration for the founder of the discipline that is unheard of in any other scientific field. Alfred North Whitehead commented that a discipline is lost if it isn't capable of forgetting its founder. Both the veneration and fear in which Freud was held have had far-reaching implications for the development of psychoanalysis.

Second, the phenomenon of scientific splits is connected to the real and apparent war that psychoanalysis has waged for its place in the family of the sciences. Scientific and pseudo-scientific objections to psychoanalysis were raised from the start and are familiar; we will not return to them here in any detail. As a man of the natural sciences, Freud did everything in his power to show that psychoanalysis is a natural science in every respect, even when doing so was inappropriate. He sought to place clear limits to the discipline in order to prevent the blurring of its attributes, and in order to prevent charlatans from mixing the dross of base metals with the gold of psychoanalysis. The limits he established were, it would seem, clear and fair: any research that both recognizes the unconscious, the phenomenon of transference and resis-

< PSYCHOANALYTIC HISTORIOGRAPHY >

tance to therapy, and also accepts the inductive method of scientific examination, can find itself within the framework of the psychoanalytic movement. However certain individuals and schools that upheld these criteria were nonetheless forced to separate themselves from the movement or were threatened with removal. Those whose positions did not suit the central stream of the movement failed to find a place in it. This process continued even after Freud's death. The splits were often brought about for insubstantial reasons and impeded the development of the field. Scientific splits occurred in other fields as well, but these were usually over professional considerations. The European Society for Physics and Chemistry, for example, split at the beginning of the twentieth century because of gains in knowledge, which required a separation for practical reasons.

Another matter that distinguishes psychoanalysis from most other disciplines is the emotional involvement in the material researched and a professional ethic that is unique to the discipline. This emotional involvement creates numerous variables that affect the scientific picture, and, as a result, psychoanalysis poses from the start a greater danger of biased treatment of the scientific material. The ethical situation prevents the free flow of clinical material that would enable colleagues to evaluate their findings. This makes it more difficult for psychoanalysts to evaluate the reliability of these findings and their relevance to the discipline's body of knowledge.

Another phenomenon unique to psychoanalysis among the sciences is the multiplicity of private theories. Private theories are scientific notions that exist in the mind of the individual analyst but do not correspond with the official theories of the day and do not come to the knowledge of the scientific community. I discuss this phenomenon in Chapter 12.

Irrational Elements in the Psychoanalytic Discipline

Up until now we have been sketching a basic outline of the psychoanalysis of science. The goal of the following chapters is to point out the existence of irrational elements in the development of the psychoanalytic discipline and to show how they influence its progress.

To this end, psychoanalytic lacunae and myths will be analysed. It is not my intention to map the unconscious factors at work in psychoanalysis in a systematic fashion; rather, I want to demonstrate the principle at work through the presentation of a few central examples. The lion's share of the book, then, deals with the analysis of lacunae. The lacunae presented here represent three categories, each one of which

< LOCATING THE IRRATIONAL, UNDERSTANDING THE REPRESSED >

derives from a different concealed motive. The lacunae of the inside of the body revolve around the difficulty of seeing and explaining phenomena within the primitive fears and anxieties they arouse. The image of the inside of the body, both its contents and its space, is a subject that is virtually underrepresented in psychoanalysis. We have very little relevant clinical material. There does not exist, therefore, any clear clinical theory concerning the inside of the body. Nor have there been serious studies of the question of what gave rise to such a central lacunae.

The post-traumatic dream as a lacuna represents the mechanism of obfuscation and blurring of a scientific problem, whose unconscious goal is to maintain the integrity of psychoanalytic theory. This mechanism highlights the difficulty of recognizing an exception that challenges the validity of the theory. As a result, the road to new discoveries becomes blocked. The avoidance of attempts to understand the mechanism and the meaning of the post-traumatic nightmare created, as this chapter shows, a chain of difficulties and delays with regard to understanding different clinical phenomena: post-traumatic neurosis, the phenomenon of trauma in general and the phenomenon of anxiety.

The third lacuna to be examined is the epistemological split of psychoanalysis, which reflects the gap between what we want to see and what we actually see. A gap exists between psychoanalytic metapsychology and clinical epistemology. Another epistemological split exists within the framework of clinical theory. I deal with the psychoanalysis of epistemology, which is to say, with the attempt to explain the motive behind the positivistic position that has held sway in psychoanalysis, despite the fact that it does not correspond with clinical experience and theory. Likewise I will demonstrate the effects of the epistemological split on the development of clinical theory of countertransference and the idea of complementarity; both are central lacunae in psychoanalysis.

Regarding the treatment of scientific myths, these have a function that goes beyond narrative. They give expression to the irrational element that drives the development of psychoanalysis. I examine the Thanatos myth and the myth of the Logos in Freudian theory. Examination of the Thanatos myth reveals the unconscious reasons that led Freud to abandon the original theory of drives. Likewise, I describe the mythological dimension of the Freudian Logos, which he sought to eliminate from the theory of drives, and I point out its link to the Thanatos myth. Freud tried to establish the Logos and grant it extraterritorial standing, while sidestepping the internal logic of his theory—all this in order to construct for the Logos, and especially the scientific Logos, a temple.

< PSYCHOANALYTIC HISTORIOGRAPHY >

Another way to expose the irrational aspect of psychoanalysis is to examine the metaphors of the scientific text. Analysis of a number of such metaphors will demonstrate Freud's emotional motives with regard to unsolved scientific problems.

< LOCATING THE IRRATIONAL, UNDERSTANDING THE REPRESSED >

PART II

The Lacuna of Images

There are two coexisting currents: the urgent need to communicate and the still more urgent need not to be found." — D. Winnicott

Part II will deal with the central lacuna of psychoanalysis: the lacuna of images of the inside of the body. Images of the inside of the body—both its space and its content—constitute for all intents and purposes blind spots in the psychoanalytic discipline (an unconscious lacuna). In Chapter 4 the scientific gaps in this area will be located as they emerge from a survey of the scientific literature. Examination of the professional literature shows that while writers have registered a sense of the repression of this subject from time to time, the problem has never been brought to the forefront of scientific discussion. Nor has speculation been offered regarding the meaning of these gaps, which themselves can be seen as a scientific symptom. Part II, then, offers a thesis that attempts to account for the phenomenon.

The repression of the meaning of menstruation blocked the option to see into the connection between the menstruation taboo and the incest taboo. Chapter 5 explores the identification of this repression, which can be considered as a scientific symptom; the understanding of its meaning facilitates the development of new scientific associations.

4

The Inside of the Body

I became interested in the inside of the body as a lacuna during my analysis of a young woman who vividly expressed her fantasies about the inside of her body.[1] When she entered analysis, she was a hyperactive woman who felt that she had been forced out of the womb and into life too quickly, and that this had made her hyper-independent during her childhood and adolescence. She was preoccupied with her skin and worried about every small injury, much in the way that small children react to their little wounds. The skin affections precipitated her decision to seek therapy.

It soon came out that the skin protected her not so much from the outside world, but mainly from her own inside, which seemed very frightening to her. She often used the word autarchy to express her wish and need for physical and psychic independence. But this did not help her much as she was convinced that everybody could see her dirt. She felt like a criminal, but she had not the slightest idea what crime she had committed. Her compulsive image for total, hermetic physical intactness had led her nowhere.

As a child, she tried to be good, but gradually she developed the tendency to torment her mother, for instance, by insisting on being allowed to see her mother's menstrual blood. Upon being refused, she told her girl friends that her mother was pregnant, which was not true.

During the first stage of analysis my room was experienced as an arena of the primal scene in which the drama was taking place inside the body of her mother. She perceived my room as a dark cellar. The curtains were full of dark, terrifying circles, monsters emerging out of a nightmare, and some of them were blood red. They reminded her of terrible crimes. She heard terrifying noises. Cats were hiding in the room that she thought might be her parents. She had to live in this room of mine—a dark and bloody inner space. This drama was a recapitulation of the primal scene with early acoustic stimulations as well as oral and anal sadistic elements.

< THE INSIDE OF THE BODY >

In the second stage of her analysis, she entered a phase of hibernation, in which her skin became the image of a womb in which she was hiding. When she became frightened, she wanted to sit under my skin 'like a leech'. When she came out of her hibernation, she started looking inside her body in order to find out what dangers might emerge from it. She became more 'flesh and blood'. She had many fantasies about falling and about abortions. Dogs fell out of windows or into open black graves. Her mother aborted her . . . The hole could never be filled in again.

She recollected her mother's refusal to show her the menstrual blood. Her mother preferred dying to showing her this blood, she said, adding that menstruation is like dying. In fact, she saw in her mother's menstrual blood the evidence of her tendency to commit suicide. 'The final abortion' was associated with her fantasies about her own imagined abortion. Every month she fantasized that she had forgotten to take the pill, and felt released when her menstrual blood appeared. Her sister was in a car accident, and this reminded her of her own delivery. After the accident, her sister was reborn. As a child my patient had fantasized the destruction of her sister, who kept on reappearing intact. She wished to believe that her sister was aborted in her mother's menstruation; then she wanted to find out whether her mother's menstrual blood contained parts of her sister. At the same time, she wanted to make sure that this was not true.

Through the repetitive imaginary act of becoming pregnant, aborting and becoming pregnant again, she expressed her murderous wish anew in order to undo it. By imagining her own abortion, she also turned the wish to murder her sister inside her mother against herself. When she referred to her mother's womb, the symbol of the vagina came up. She described it as a mouth full of teeth, comparing me with a dentist.

The demand to see her mother's menstruation expressed the wish for fusion. She talked in terms of a blood treaty. Menstruation revealed itself in the analysis as a condensation of several factors: her masochistic identification with her mother; her wish for a symbolic unification with the mother; her wish for a blood treaty between the guilty.[2] Her double wish to see the pieces of the sister in the mother's menstrual blood and to find evidence to the contrary; and, above all, the idea of the condensed life—death symbol interconnected in the menstrual blood. The different meanings of menstruation represented many of the patient's drives and conflicts. She connected menstruation with defloration, sexual intercourse, pregnancy, delivery, death, bleeding, hysterectomy, and artificial and natural abortion—and at the same time associated all of these with the mother's inner space, which was filled with babies and blood. George Devereux found similar associations in his study of the

< The Lacuna of Images >

Mohave Indians.[3] The two most important meanings in the primary process productions were: (1) the wish to share with her mother either the guilt or the innocence concerning the lot of her sister; and (2) the frightening idea that menstruation means life and death at one and the same time. When the mother loses parts of herself, is it the beginning of the end, or does it mean new life? Perhaps it is both at the same time? When my patient referred to the menopause of her mother, she felt that she herself, with her own menstruation, was living on account of her mother, who was going to die.

Menstruation meant abortion, delivery and dying at the same time. The abortive element of menstruation was the link between content and space. When her intactness was threatened, the content might have fallen out into space. Falling meant losing the familiar womb, falling apart, crashing and falling into dark holes; it meant losing parts of her body. Falling in space had the quality of dying, of annihilation.

Coming to life meant dying at the same time as well. My patient's fear of being aborted, of falling, was not just guilt feelings about her aggressive fantasies. They also represented her very basic, primitive, primary, psycho-biological fear of being dropped. She spoke in terms of clinging or sticking when she wished to stay under my skin to escape the fear of being dropped.

In my patient's description, the inside of her body was undifferentiated. The reproductive urinary and gastrointestinal systems were interchangeable. Her vivid and plastic fantasies about the inside of her body and that of her mother expressed, concretely and metaphorically, the fantasy of the most basic form of introspection. Here I wish to stress that my patient was a 'typically' neurotic person and not borderline. She had effective defences and her ability to see inside herself was not a result of loose ego boundaries.

During the analysis of this patient it became clear that such vivid and intensive fantasies of the inside of one's body are quite rare. It never came up in any of my other patients, and I had not heard much about it from my colleagues. When I searched the psychoanalytic literature, I was surprised to discover how little had been written about these important aspects of the body.

While Réne Spitz views the inside of the body as a primary source for the feelings,[4] and Marion Milner describes it as a new womb that replaces the external womb of the mother—a place that provides the basic security that enables one to step into life—the psychoanalytic literature seems to view this interior more as a source of fear that does not tend to reveal itself easily.[5] Our main and most important source is Melanie Klein, who based her ideas about the inside of the body on the analyses of children aged 2 and a half years to four and a half years.

< THE INSIDE OF THE BODY >

Klein found that small girls fantasized about the embryos, faeces and blood that fill the inner space of the mother; they enact their imaginary penis, which they take from the father, to destroy mother from the inside. "The child believes that it is the body of her mother which contains all that is desired."[7] When they look at their menstrual blood, women have the notion that the children inside them are damaged or destroyed. In contrast to little girls, little boys attain a feeling of the sadistic omnipotence of the penis, which enables them to master their early anxieties through externalization. Girls, on the other hand, remain under the rule of a hidden and unknown inner world.

Judith Kestenberg stresses the connection of femininity to bleeding holes. My patient tried all the time to contain the frightening elements inside her.[8] Devereux reached similar conclusions from the anthropological point of view. He posited that the tendency to perceive the body as a shell or solid matter is a reaction formation to repressed childish sadomasochistic fantasies concerning the contents of mother's inside. Devereux succeeded in connecting anal, genital, vaginal and oral defloration, abortion, delivery and menstruation. In his view, their common denominator is the child's fantasy to remove the inner content of the mother from the internal cavity.[9] (My patient seemed to represent Heckel's biogenetic rule: she recapitulated in a condensed manner the beliefs and rituals of various tribes.)

In trying to find out about children's image of the inside of the body, Paul Schilder asked children what they believe they have in inside them. The classical answer was 'food.' Children prefer to see the inside of their bodies in terms of what they put into it. This gives them the feeling of mastering. As Schilder explains: "It is one of the paradoxes of our bodily experience that our sensations relate to the surface of the body, and yet we do not regard the actual surface as our body proper . . . we are inside our skin and know nothing directly of the interior of our bodies."[10]

Erik Erikson used children's games to analyse the different uses that girls and boys make of space. He found that girls generally refer more to inner space, while boys invest their interest in outer space, that is, that girls are busy with the interior of the house and its arrangement, while boys tend to play with towers etc.[11] Here we should remember that Freud called the house the symbol of the human body. We can therefore deduce that women are more preoccupied with their bodies than men.

The inner space leaves the women with a feeling of loneliness and of being empty and dry—robbed of her treasures. At the same time, it symbolizes loneliness and the fulfilment of wishes. With every menstruation the feeling of mourning unborn children returns, and with the arrival of menopause the scar becomes irreversible.

Enid Balint is one of the few analysts to represent an analytic case in

< THE LACUNA OF IMAGES >

which the inside of the body of a grown woman is pictured in a concrete way.[12] She connects the sense of inner emptiness with the feeling that the inside of the body is filled with dead and unworthy trash, like the straw with which puppets are stuffed. The patient that she describes fantasized that she was full of articles and dead people. A scan of the psychoanalytic literature indicates that fantasies about the inner space and its contents are an extremely neglected chapter. On the basis of his clinical experience, Owen Renik concluded that women do not generally talk about their menstruation in analysis on their own initiative, and certainly not about their fantasies regarding it. When encouraged by their analyst to do so, they speak mainly about symptoms connected with menstruation and not about their fantasies connected to it.[13] This is also my personal clinical experience.

Mary Chadwick claims that the seclusion of women during menstruation throughout cultural history was brought about by women themselves in order to protect the community from infection. Menstruation taboos became a historical necessity, as the menstrual period disturbed the social order of primitive societies; women therefore had to seclude themselves during menstruation in order to protect themselves and society from their anger, vengefulness and jealousy.[14] In analogous fashion, witches threatened society during certain periods. A witch does not give birth, but rather aborts and sacrifices children in bloody rituals.[15] "The monthly neurosis", as Chadwick calls menstruation, universally gives expression to feelings of vengefulness and jealousy over all things masculine. It also gives rise to an inclination to steal and harm, and to suicidal tendencies.[16]

Mary Lupton concludes her comprehensive book on menstruation and psychoanalysis by claiming that the menstruation taboo is considered the most malignant of all taboos. In the context of the psychoanalytic discipline too there exists a tabooization of the menstrual period. Menstruation is repressed in the framework of the psychoanalytic discussion.[17]

Although there exists an unspoken agreement between mother and daughter that menstruation belongs solely to the mother, analysis of my patient showed that conflict erupts when the daughter discovers the existence of the blood-pact between them. The pact requires collaboration on this subject, and the mother violates the pact by claiming it all for herself. The blood-pact gives expression to the common guilt with regard to the killing of the child (as with the blood-pact around the killing of the father in *Totem and Taboo*). The alternative is to prove common innocence. Of all cases of fantasies involving blood, menstruation arouses the most murderous fantasies. The blood pours forth from the sexual organs of the mother. It contains blood and blood clots, which

< THE INSIDE OF THE BODY >

can be seen as the remains of the embryo. In this context it is possible to explain the menstruation taboo as an attempt to conceal the *corpus delicti*. Devereux writes that menstruation as a psychoanalytic subject is a virtual *terra incognita*. In describing the Mohave women, the subject of his psycho-anthropological research, Devereux points out that, whereas they are very open when talking about their sexual experiences, they become quite reserved when it comes to the subject of the reproductive system and all its ramifications.[18]

Eva Lester and Malkah Notman investigated the psychodynamics of pregnancy and commented on the dearth of papers on the subject.[19] And, in his work on multiple induced abortions, Newell Fisher expresses his surprise about how little psychoanalytic investigation there has been concerning the psychology of unwanted pregnancy. He attributes this to an active wish among psychoanalysts not to think about the subject.[20] Devereux says that menstruation as a subject of psychoanalysis is terra incognita.[21] In his paper on the inside of the body, Howard Steward comments on his finding almost no relevant literature on the subject.[22]

Erik Erikson goes even further. When he wrote about womanhood and inner space, he concluded as follows: "Clinically this void is so obvious that generations of clinicians must have had a special reason for not focusing on it."[23] However, like other psychoanalysts who felt that something strange was going on, he never went beyond pointing this out.

Frieda Fromm-Reichmann deals briefly with the tendency of analytically oriented therapists to avoid psychotherapy with pregnant women. Their rationale is that the cathexis of the pregnant woman is directed mainly toward their inside (which they do not express in therapy). Fromm-Reichmann, on the other hand, is in favour of proceeding with intensive therapy during pregnancy.[24] But the obvious question is: Why does the psychoanalytic community resist relating to the inside of the body or to fantasies about the developing foetus? Until now no one seems to have asked this question. Nor has anyone asked why therapists accept this attitude so casually and seem to take it for granted. For it is clear that quite a number of women must have continued their analysis through at least part of their pregnancy, yet this is not reflected in the psychoanalytic literature. There is probably nobody in psychoanalytic literature who dealt more with the psychodynamics of pregnancy and delivery than Joan Raphael-Leff. Even though she studied many pregnant women, she seems to feel that the mystery of the inside is insolvable, and she says that "the eternal essence of pregnancy lies in the sensation of strange, disturbing, uncontainable emotions which must be contained because they cannot be shared,

< THE LACUNA OF IMAGES >

expelled or vaporised without loss".[25] Even though it is obvious that Raphael-Leff is extremely sensitive to the inner experience of women, there are very few concrete descriptions of the inside of the body in her work. The following passage is representative: "When they puncture my skin, I feel I might come pouring out as if there is a vacuum inside and my inner self is liquid—as soft as the yolk of the soft-boiled egg which I have never eaten". Her descriptions concerning the inside of the body of the pregnant woman seem to be more about the experience than the experience itself. Often, just when one expects a description, one gets a semi-abstract text followed by a clinical description (of symptoms). For instance: "Continuity of self is disrupted by internal disturbances which disturb her ordinary illusions of unified identity and indivisibility. Pregnancy throws into question body boundaries which since babyhood have defined the separateness of her own self within her own skin."[26]

In thinking about the difficulty in fantasizing about the inside of one's own body in general, it occurred to me that psychotic and borderline patients might be able to offer more associations in this direction because of their weak defence mechanisms. For instance, we know from the psychiatric clinical pictures of the manic depressives that, during their psychotic-depressive state, they ruminate about changes in their inside. We also know from their psychotic pictures about their phobias regarding the inside of the body. I therefore began to search for more concrete descriptions of the inside in textbooks with psychodynamic orientation, such as those of John Frosch[27] and Thomas Freeman,[28] but without any success. I then looked for such descriptions in psychoanalytic case studies of psychotics and borderlines as described by Harold Seals[29] and Herbert Rosenfeld, [30] Heinz Kohut[31] and Otto Kernberg,[32] where empty spaces, fragmentation and the like are described. Since analytic work with this type of patient is connected with a blurred distinction between the inside and the outside, it seemed that we could expect to learn more about the inside from them, especially because psychotics have a tendency toward concretization. However, strangely enough, the literature on this category of patients contains almost no concrete material about the inside. In those instances where there is some sort of concrete description, authors quickly shift to the abstract. Thus, in reporting on a psychotic patient, Harold Searls writes: "She felt she was swelling like a balloon. It was unpleasant. Expectancy had something to do with it. If she expected something from another person or from herself, this symptom increased."[33] Self-psychologists, who are credited with a strong and genuine interest in the inner universe of the individual, have not been any more successful in demonstrating this concrete universe. David Rosenfeld comes closest to reaching descrip-

< THE INSIDE OF THE BODY >

tions of the inner space during a psychotic stage. Here is how he describes the fantasies of a psychotic girl: "The boundaries of her body were not the skin, the muscles, the skeleton, but only arterial walls which play the role of the external skin and which become empty of blood when there is an anxiety crisis and lack of contention." In Rosenfeld's analysis a weak arterial wall is typical of a psychotic-primitive body image. Such patients have lost the psychological notion of skin and feel that their body scheme is a vital fluid, sometimes conceptualized as blood, contained by a weak membrane or wall.[34]

In seeking to discover what interest psychoanalysis has taken in the contents of the inside of the body, I asked myself how psychoanalysis has approached the problem of menstruation and of blood in general. Some of the questions I sought to answer were: What do we mean by the instinctive fear of blood, and at what developmental stage is it said to occur? Is the fear of blood in general as strong as the fear of menstrual blood? However, the psychoanalytic literature has not concerned itself with either the dynamic meaning of blood or the fear of blood in general, or with the fear of menstrual blood.

Jannine Chasseguet-Smirgel has a very simple explanation for this striking omission:

> We find substitutes for the body and parts of the body of the object, thereby protecting the latter from our direct attack. Conversely, the blood is very seldom symbolised, as if it represented a *non plus ultra*, for which we are unable to find unconscious equivalents—at most we may discover allegorical representation, but not true symbols in the psychoanalytic sense.[35]

If Chasseguet-Smirgel is correct in her hypotheses, this would satisfactorily explain why psychoanalysis has not taken an interest in blood. However, I do not agree with her. Blood symbolizes the soul in many cultures. It is not solely the source of physical life. If blood is not a true symbol with unconscious significance, why does it play such a central and various roles in almost all cultures. Why are so many taboos connected with blood?

In dealing with taboos and the fear of blood, Freud never made a clear distinction between menstrual blood and other blood. In "The Taboo of Virginity", he speaks of the "horror of blood among primitive races that consider blood as the seat of life. The blood of defloration certainly is no menstrual blood and still is a source of horror." He blurs the distinction even further in stating: "The taboo of virginity is connected with the taboo of menstruation."[36] In *Totem and Taboo*, he had stated that the totem has the same blood as the human being. Therefore, the blood ban—in reference to defloration and menstruation—forbids sexual intercourse with a woman of the same totem.[37] Devereux comments:

< THE LACUNA OF IMAGES >

The fact that this statement couples defloration which is in fact a result of 'aggression' with menstruation, which biologically is not caused by aggression, clearly implies that the later phenomenon is unconsciously viewed not merely as a result of menstruation (damaged genitals) but also of fantasised aggression.[38] This, of course, strangely supports the thesis that all forms of genital bleeding are, at least unconsciously, imagined to be the result of aggression.[38]

From that point on, blood as such is hardly discussed in the literature, except by Devereux, and when it is, it deals only with menstruation. Helene Deutsch contributed to our knowledge of the psychodynamics of menstruation, stressing the different attitudes of girls to their menstruation in different developmental stages.[39] The finding of Deutsch that is relevant to our discussion is that women tend to hide their menstruation as their most discreet secret. They may discuss many intimate matters with their daughters, but not this, which they prefer not to share.

According to Deutsch, if a woman were on trial for murder and could prove her innocence by showing that the blood in question was menstrual blood, she would rather convict herself than say this. (Deutsch wrote this many years ago and it may seem exaggerated to us today, but it does describe the most intimate feelings about menstruation that still come up in analysis.) Elsewhere she mentions that the frightful image of the bleeding sexual organ relates to the internal organs as well. The frightening nearness between living and dying in these biological functions of the woman overclouds even the danger of feminine castration.[40]

Although there is a silent agreement between mother and daughter that the mother's menstruation belongs only to her, analysis with my patient indicates that conflicts arise when the daughter insists on a 'blood-pact' with her mother and the mother refuses. The blood-pact represents the shared guilt concerning infanticide (which is similar to the shared patricide among males in Totem and Taboo). The alternative is the proof for shared innocence. Menstruation awakens the most extreme murderous fantasies as compared with other blood. It emerges from the mother's genitals and contains blood clots and debris which may be fantasized as the remains of a foetus. In this context, the menstrual taboo can be interpreted as an effort to hide the corpus delicti.

For many years the Oedipal complex and fear of castration took up much space in psychoanalysis at the expense of earlier fears of disintegration, falling into space and purification. These fears seemed connected with the sensation of death. With the exception of Melanie Klein, menstrual fears were not connected with these motives. Even in

< THE INSIDE OF THE BODY >

Klein, we find only a few remarks about menstruation fantasies in children. According to her, menstruation embodies the girl's fear that her body will be attacked by the mother, whether because she wants to get hold of the father's penis or because of her sadistic fantasies during menstruation. Girls also fear being attacked by internal objects. Menstrual blood provides them with evidence that children inside the mother have been damaged or killed.[41] Klein does not take into account the sadistic role of the girl, which is that she herself might be the murderer of children. Thus, in contradistinction to the diffuse image of the inner space, menstruation is a very realistic and concrete factor and explains why the repression of these fantasies may be so strong. It may explain the disappearance of murderous fantasies concerning menstruation, when such fantasies are so vividly manifested in the context of the inner space.

That menstrual blood is the only direct and concrete proof of the terrible things that go on inside the body in itself grants menstruation a unique standing. Menstruation is sometimes associated in analysis with pregnancy, abortion and birth. My analysands associated with super-menstruation[42] an image of pregnancy that is also held among Mohave women.[43]

In the literature, we sometimes come across the image of menstruation as an abortion of a fantasized pregnancy, or as the murder of an imaginary foetus. Sometimes, as in Helene Deutsch, we read about death in association with "giving birth to menstrual blood".[44] The image of menstruation as connected with pregnancy, on the one hand, and with abortion and death, on the other hand, is perceived by women as an intermediary between life and death. The most vivid interest in the psychoanalysis of menstruation, however, was shown by an anthropologist.[45] Claude Dagmar Daly reached his conclusions on the basis of his personal analysis and supported them with anthropological material. In his view, 'the menstruation complex' is the number one taboo among humans and was created by the male. The value of Daly's work lies, on the one hand, in its discovery of the significance of the male in the riddle of menstruation, and, on the other hand, in trying to bring menstruation to the centre of psychoanalytic discussion. In his view, the female sexual libido peaks during menstruation. Thus, castration fears and even earlier male fears resulted in an inhibition of sexual desire and foreplay, which decisively inhibited the sexual development of the human race. Among the questions that he posed were: Why didn't psychoanalysis explore menstruation? Is the "nonchalance and ambivalence concerning my theory" indicative of something.[46] He suggested that, in claiming psychoanalysis is interested in infantile sexuality but not in later development, psychoanalysts were rationalizing their inability to come to

< The Lacuna of Images >

grips with the subject. In his view, many phenomenon that were explained through female envy of the male could more profitably have been explored in terms of narcissism and fear on the part of males. By refusing to do this, male analysts were able to keep from exploring an unsettling subject. Daly claimed, therefore, that psychoanalysis refused to deal with matters that endanger the male narcissistic position by bringing to light fears of the female.

Impressed by Daly's work, Sandor Ferenczi declared that anxieties connected with menstruation are no less central than castration fears.[47] His enthusiasm about this area of inquiry resulted in a doubling of the number of researchers on the subject at the Centrale Zeitung für psycho-analytische Pädagogik.[48] A sort of 'Menstruation Festival' ensued, with contributions by Melitta Schmiedeberg, Karen Horney and others, but no breakthroughs resulted. Although the psychoanalytic community allowed Daly to present his ideas, they did not follow his presentation with serious scientific discussion. Thus, his ideas remained a foreign body within the body of psychoanalytic knowledge, and the subject again fell into oblivion for many years.

Avoidance of the subject continued for more than forty years, until Ruth and Theodor Lidz commented that:

> The theory of Daly has not set off resonance in others because of the intensity of the repression. We are not in the position to clarify the understanding of this very important fear which, if not universal, is certainly extremely widespread, and thus must reflect something rather fundamental about human behaviour.[49]

The Lidzes made some very interesting observations, but, like Daly, they were mainly anthropological in nature. Investigation of the male menstruation phenomenon in Papua New Guinea resulted in the conclusion that Papuan males envy females for their reproductive potential and identify with them. Where women purify themselves through menstruation, the men make their noses bleed. Simultaneously with this envy, Papuan males also fear menstruation. They believe that menstrual blood entering their urethra may make them pregnant and consequently doom them to death together with the foetus, because there is no way for the foetus to get out. In order to keep from being endangered, men live separately from women. Like Daly, the Lidzes connected the male to the 'menstruation complex'.

With the above in mind, one is forced to ask whether the denial of menstruation and the inside of the body is an exclusively male narcis-sistic defence, or whether men and women share the fear of touching on the painful and frightening problems that concern them both.

The inside of the body concerns not just its contents, but space and falling into space as well. Here, too, we witness a lack of psychoanalytic

< THE INSIDE OF THE BODY >

interest. We have hardly any concrete psychoanalytic descriptions of falling, falling apart, disintegration and the like.

The void concerning one of the central motifs of mankind in psycho-analysis—the inside of the body and the menstruation-pregnancy-parturition triad—is doubly strange considering the centrality of these themes in human life. We include all derivatives of the contents of the inner space, including all types of bleeding, such as genital, menstrual, defloration, abortion, as well as many other phenomena in this triad, within the framework of psychogynaecology. If we look at the place of this triad in anthropology and in the history of religion, we find that, in many cultures, the inside of women's bodies symbolizes pregnancy, abortion and death at the same time. Female goddesses often symbolize the epitome of life and death. During menstruation, women are God's property. Even more, the great totemic substance inhabits the menstrual blood.

The symbols of the great mother are mainly derived from the inside of the body image that symbolizes monsters, death and destruction. According to Daly, the Indian goddess Kali represents the most shocking instance of the terrible mother.[50] In cultures where the inside of the body is said to contain the soul, the inside of the body becomes a somatic symbol for the unconscious.

Menstruation and blood play a central role in the history of culture, and in almost every culture there are numerous institutions and laws connected with both topics. Likewise there are endless mythologies reflecting menstruation as the simultaneous source of life and death. The phenomenon is so all-pervasive that one can safely say that the menstruation taboo has an almost universal dimension, one beyond that of even the incest taboo. In fact, it is the most universal taboo. According to Robert Briffault, all taboos were derived from the menstrual taboo.[51] Menstruation is both the elixir of life and the source of death. Blood in general is a central motif in the history of religion, mainly in connection with blood sacrifice. Blood was a leitmotif among the Aztecs, who, despite their highly developed culture, had the bloodiest rituals of all. The sacrifice of blood to the gods was meant to prevent the eradication of the population, and in general the constant supply of blood was meant to pacify the gods and prevent punishment. All of the Aztecs' notorious war initiatives were motivated by their desire to obtain as many captives as possible for sacrifice. The cruelty of this culture reflected its extreme anxiety *vis-à-vis* nature's threats. Here we see how the close connection between life and death is symbolized by blood. After all, the Aztecs gained life by sacrificing life.

It would therefore seem that the inside of the body and the preg-nancy–parturition–menstruation triad occupies a central place in all

< THE LACUNA OF IMAGES >

human cultures, but this is not reflected in psychoanalysis. Yet the mystery connected with the inside of the body, which throughout history has had such a frightening effect on humanity, is not a total stranger to psychoanalysis, for the menstruation taboo has encroached itself on both analysands and psychoanalysts—who have also undergone analysis. Some archaic inhibitions, however, seem to prevent a thorough analytic exploration of the phenomenon, resulting in an unconscious pact between patient and analyst to leave the subject largely unexplored. Jean Bertrand Pontalis notes: "In order to prevent the threat of internal destruction, dangerous internal objects must be placed outside themselves, and reality must be used to maintain the frontier of the ego . . ."[52] The same would seem to be true of our somatic ego. We seem to project deep early anxieties about our inside to the outside and to repress our capacity to relive them.

Naturally, this process is also reflected on the theoretical level. We should not forget, for instance, that what Melanie Klein came to know about the inside of the body was based mainly on projections of the child on his/her mother. Similar to the menstruation complex of men, which keeps women at a distance, psychoanalysis keeps the problem at theoretical arm's length. For dozens of years the Oedipus complex and the fear of castration took up space in psychoanalysis at the cost of discussion of earlier fears, such as disintegration, crushing emptiness, falling in space, purification, fears that come much closer to the image of death and non-existence. Only after psychoanalysis began to deal with psychotics and borderlines, and with the rise of self-psychology, did the awareness of earlier fears and images in adults receive greater attention. Of course, it is easier to get productions of earlier fears and images from patients with severe defects, but even in such cases we see that the capacity to produce associations about the inside is limited. And, if it is so difficult to produce concrete associations regarding the inside of the body in those whose defence mechanisms have failed, it is even more difficult in neurotic patients, who bring these matters up in a more disguised fashion, if at all. Both analyst and patient find it easier to ignore these motifs and let them fall back into the unknown.

It is interesting that most of the work done in the realm of our subject has been anthropological in nature, but despite the importance of anthropological material to the validation of psychoanalytic observations, psycho-anthropology can function as a defence. Why, for instance, were psychoanalysts like Bruno Bettelheim, Géza Roheim and Theodor Lidz unable to demonstrate their cases with clinical material from the analytic room? What drove them all the way to the Papuans instead of seeking an answer a metre away, on the couch? Anthropology facilitates taking flight to far-off lands where we can avoid experiences

< THE INSIDE OF THE BODY >

of the self. Anthropology should in fact assist us in revalidating psycho-analytic material, but it should not take its place.

Menstruation symbolizes the great power that women have over life and death. Woman is directly responsible for the preservation of life, and it is up to her whether she allows the foetus to develop or destroys it. Like pregnancy, menstruation represents a primary fear to both males and females, for the uterus is a double symbol (Janus faced in Erikson's words): it combines both the preserving, protecting aspects of the female and her ability to expel and abort. While men can try to defend themselves against their fear of this double power by distancing them-selves from women, women cannot similarly employ such a distancing defence. Menstruation is a steady and recurrent reminder of a most basic fear, and women can only repress their fantasies and keep them from entering the conscious.

Another element of the triad taboo, pregnancy, can be threatening, especially on the level of fantasy. The female fear of damaging her foetus, and even of killing it inside her, comes to its climax during the post-partum psychoses. This not only means infanticide, but goes even further. The foetus is biologically (and psychologically) part of its mother, and she identifies with it—having once been a foetus. The fantasies, especially the unconscious ones, that she might inflict death on her embryo produce in her the danger of self-extinction. Devereux researched artificial abortions in different cultures.[53] In some areas, suicide was considered a technique of abortion when other methods failed. In one of the techniques of aborting that he mentions—jumping from a high place—we can find the connection between abortion and self-abortion, for example, suicide. The mother's fantasies that her child will be born with a congenital defect (a frequent fantasy) are connected with the fear of her own extinction on the most basic physical level. This indicates that the fear is much earlier developmentally than superego fears. (The early anxiety fears of Melanie Klein are already superego fears.) These fears are connected to the most primary sensations of the self.

The fantasy of getting pregnant, aborting and getting pregnant again in an endless cycle—as was demonstrated by my patient—represents the life–death cycle. This is a play on the border between living and extinction. Again, this is not just a repetitive infanticidal fantasy, but playing with one's own life. Although I have the feeling that this is quite a common fantasy, it is another undiscovered motif in psychoanalysis. Dinora Pines discusses eight different motifs connected with the cycle of repetitive abortion, including omnipotence, regression, Oedipal wishes and pathological mourning.[54] Pines does not, however, describe the most basic fear—that of the self.

< The Lacuna of Images >

Returning to fears connected with space, my patient described her falling into space and out of the womb (her birth-fantasy). This fear has nothing to do with object relations, as described by Klein. It is the most basic psychobiological sensation of being dropped. In order to sense the feeling of falling, disintegrating or crushing, one does not need the family triangle. When my patient told me that she was under my skin, she was speaking in terms of clinging or sticking. It was her reflex against falling out of her mother or against being dropped.

In referring to the skin in the light of early object relations, Esther Bick speaks in terms of adhesive identification—a way of holding oneself together by attachment to an object in which there is no projection or introjection, only 'sticking'. This sticking is the psychological equivalent of the grasping reflex of the ape (and the infant) that prevents it from dropping or being dropped.[55]

Falling in space is a source of primal fears. The Hebrew word for space, chalal, is also used in connection with death. The famous horror vacui is a projection of this internal fear of being dropped, which has its origins in the inner space. This brings up the question whether there are limits to human fantasy, and if so what they are. In modern physics the speed of light was found to represent an absolute limit. The absolute limit of the human imagination is death. That one cannot imagine oneself dead can be demonstrated by a simple thought exercise. Whenever a person fantasizes about being dead, he must of necessity view himself as a non-object. But imagining oneself as a non-object would be manipulating oneself into an ontological confusion. This is the limit that one cannot pass. Death can only be defined in terms of fantasy in negative terms, that is, it is the only thing that is unimaginable. Thus, looking into the inside of the body, even in fantasy, is an extremely alarming sensation. It would seem to provoke a feeling of proximity to the zero-point of the human fantasy. The inside of the body associates in different ways with this closeness between life and death, which is a sensation that must be prevented at all costs. Therefore, it is not difficult to see why psychoanalysis has insisted that castration fears are synonymous with death fears. It is done in order to dull the fear of death. As such, it is an expression of the negation of death.

The fear of returning to the self as it is reflected in the history of psychoanalysis is the fear of the proximity of cessation. Paradoxically, the closer one gets to one's basic sensations, the more one approaches the sensation of one's death. Thus, distancing oneself from the experience of the self is derived from the basic rule of the limits of fantasy.

The ontological confusion concerning one's death should not be confused with suicidal features and death fantasies. Even when a person commits suicide, that person denies his or her death. A woman's

< The Inside of the Body >

confrontation with the death of an embryo represents the confrontation with death in an almost concrete way. Because the embryo is a part of herself, it is the closest she ever gets to the fantasy of her own death. When a woman imagines her body full of menstruating blood, she returns to the fantasy of the death of her foetus and of herself. This connection may help explain the enormous difficulty in relating to the inside of the body. Still, as my patient and other patients have proved, it is possible, although often with the help of projections. We do not know what enables some people to do it, nor do we know how much the ability to do so depends on the analyst.

In summary, the fear and avoidance of the inside of the body is reflected in psychoanalysis as almost a complete lacuna, while the menstruation-pregnancy-partition triad that represents the inner space in culture is the cornerstone of the content of the inner space. These three elements and their derivatives are lacking in psychoanalysis. The inner space remains a huge problem to be sensed, analysed and discussed. It is denied by both men and women, but because the male is anatomically more protected, he projects his fear of the inside on the female who is more biologically exposed to her fears. This is reflected in psycho-analysis via the way in which the incest is discussed in terms of male psychology and menstruation as a female problem—a dichotomy that has made the possibility of getting into this emotionally loaded ques-tion even more difficult because it does not take the full dynamics of men's fears in this connection into consideration.

The fear of blood is a confused and almost thoroughly denied subject. It is not fear of blood as such, because male blood, for instance, in circumcision, is not feared; nor is it exclusively menstrual blood, as shown in the taboo against virginity. The fear involves female blood exclusively, blood that comes from the genitalia and represents symbols on the border of life and death. Being so close to the idea of self-annihi-lation, these symbols are very difficult to fantasize. Thus, one of the most central issues of human life remains almost completely disguised and hidden from clinical view.

I raise the hypothesis that we have so little analytic material on this subject because both analysts and patients have entered into a pact to keep it hidden.

< THE LACUNA OF IMAGES >

5

Incest and Menstruation Taboos

Taboo does more than express the self. It constitutes the self.
— Alfred Gell

Repression of the problem of menstruation and the inside of the body has blocked our ability to examine whether there is a connection between the taboo of menstruation and the incest taboo. In this chapter I shall endeavour to demonstrate that such a connection exists, and that detailed examination of the connection is crucial for the development of psychoanalytical understanding.

The meaning of the incest taboo and exogamy rules have occupied us for the last hundred years, since the first theories on these matters made their appearance on the scientific stage. Anthropologists, sociologists, psychoanalysts and biologists have tried to illuminate the subject from every possible angle, but whenever a phenomenon is explained by many different theories, we may generally conclude that we are, for all practical purposes, still in the dark. And this does seem to be the case with the incest taboo. One of the obstacles in the path of our exploration of the subject is the debate about the universality of culture. The two poles in the debate are represented by Franz Boas, according to whom we are not in a position to deduce things from one culture to another,[1] and George Devereux, who claims in his *Elementargedanken* theory that the major cultural phenomena are universal,[2] and that motifs which appear manifestly in some societies exist in a repressed form in others.

The incest taboo and the menstruation taboo exist manifestly throughout almost every society in the world. Whenever the incest taboo does not appear to exist, for example, in highly civilized societies like ancient Egypt, Cambodia, Peru and Russia,[3] this is due to the prerogative of kings and the ruling classes. Thus, James Frazer's "divine" kings all had ritual incest in common, particularly with their mothers. But because they were superhuman, they were not struck by lightning when they trespassed this taboo. They are the exceptions that prove the

< INCEST AND MENSTRUATION TABOOS >

rule. Thus, it would appear that even the most extreme separatists of the Boaz school would find it difficult to explain why at least the incest taboo should not be discussed in terms of a universal phenomenon.

Another obstacle in our path is the confusion between the incest taboo and endogamy. Incest refers to sexual relationships within the family, while the endogamy relates to intra-familiar marriage. As we know, some societies prohibit endogamy and tolerate incest, but not vice versa.[4] A third obstacle to our understanding of the incest taboo lies in the differences in marriage laws between patrilinear and matrilinear systems—differences that are further complicated by the variability within each of these systems insofar as laws of incest are concerned.

Despite the fact that no theory, biological, sociological or psychoanalytic, has led to a breakthrough in our understanding of the incest taboo and the taboo on endogamy—the manifold theories on the subject can shed light on the structure of individuals and societies. It is in this context that I shall try to illustrate how an anachronistic anthropological theory of Emile Durkheim[5] that has been cast aside may, with some modification and when viewed in a new light, lend fresh meaning to the incest taboos and endogamy. The main point here is not to prove the correctness of this theory, but to utilize its premises to help us understand human structures, and then to show why the theory was avoided in psychoanalysis.

Joseph Shepher categorized theories that endeavoured to explain the incest taboo into three schools:[6] the first school he calls the *family socialization school*. This group includes Freud,[7] Bronislaw Malinowski,[8] Brenda Zeligman,[9] George Murdoch,[10] and Talcott Parsons.[11] While each of these represents a different theory of human culture and organization, they all believe that the incest taboo has preserved the unity of the family and thereby fostered individual development. Indeed, Malinowski insists that, without this taboo, the institution of the family would collapse.[12] The second school is the *alliance school*. Claude Lévi-Strauss is a classical representative of this school, which relates exclusively to large social groups. Its adherents claim that incest rules came into existence to bind the nuclear family to broader social units; this made it possible for political and economic relationships to take hold between families. Lévi-Strauss claimed that it is the role of culture to ensure the existence of the group. In this context, Lévi-Strauss views exogamy as a means of economic exchange: women for goods.[13] Another adherent of the alliance school is Edmund Taylor, according to whom endogamy is a policy of isolation. It is a matter of political alternatives between marrying out and being ruled out.[14] The *sociobiological school* is the third group that Shepher distinguished. The main representative of this school is Eduard Westermark.[15] He and its other

< THE LACUNA OF IMAGES >

proponents believe that people would refrain from incest even if it were not prohibited because there is a biological mechanism that prevents it. This school denies the need to explain the taboo on social grounds, arguing that there is no more incest among other mammals than among humans. Incest is a question of inhibition, not prohibition. Incest among kin is instinctively avoided because of proximity rather than kinship. Westermark believed that even without the widespread social taboos, incest would hardly ever occur.

The theory I represent falls within the boundaries of the alliance school. In 1898, the journal *Anneé sociologique* published a paper by the French sociologist Emile Durkheim, one of the pioneers of research on the incest taboo. This paper, "The Incest Taboos and Its Origins", was the first attempt to connect incest and menstruation. Durkheim describes a society in which there is total identity between the clan and its totem. Everybody is the totem and the totem is everyone. As the unifying element of the group, the totem dwells in every one of its members. In order to preserve the unity of the group, every member must remain intact. Members of the clan are one family, one flesh, one blood. The totem dwells in the blood of every member. The supreme force exists in the blood of every individual, and as the blood runs out of the body, the supreme force leaves along with it. Blood is the source of magico-biological union among clan members. The loss of individual blood means the annihilation not only of the individual but of the collective soul as well. The fear of losing the blood is expressed in particular in relationship to menstruation. This loss of female blood is critical because women are responsible for tribal continuity within the matrilinear system that he was studying. Women lose blood regularly, but the fear and respect that they command stem not from the fact of menstruation as such, but from menstruation involving the loss of blood. Durkheim connects all the magical beliefs and prohibitions associated with menstruation to this loss of blood/soul.

The taboo against endogamy is meant to prevent the combining of blood from close relatives, as this represents both the tribal totem and the family. However, since the totem of 'the other' tribe carries no magical meaning, there is no danger in mixing blood with a different tribe. Durkheim's theory therefore implies that the incest taboo is nothing more than the rudiment of exogamy. In exogamous marriages, the prohibitions related to the female, which reflect the fear of menstrual blood as an instance of the fear of blood, does not exist, for the fear of a common totem being violated is put to rest in such marriages.[16]

Lévi-Strauss, a fellow member of the alliance school, criticizes Durkheim on several points. First, Lévi-Strauss contends that

< INCEST AND MENSTRUATION TABOOS >

Durkheim's theory is based on universalizing facts collected from a limited number of societies. Like Freud and Geza Roheim, Durkheim believed that the Aborigines of Australia are the most ancient tribes on earth and could therefore serve as archetypes for primitive societies. Lévi-Strauss argues that they do not necessarily represent societies elsewhere. The theory of Durkheim explains incest as an extreme conclusion of the exogamy laws. Lévi-Strauss also claims that Durkheim did not succeed in explaining the transition of the human mind from the totemic belief to the fear of blood, from the fear of blood to the fear of women, or from the fear of women to exogamy laws. In many societies the totem was eaten after the performance of rituals. Moreover, the fact that prohibitions against contact with a menstruating female are imposed on exogamous relations in the group indicates that the blood of the 'others' has no different meaning.[17]

I would add additional reservations to those of Lévi-Strauss, who I believe did not go far enough. Durkheim explains the uniqueness of women's blood in relation to the matriarchal system, but does not explain it in relation to patriarchal systems. But the main—and perhaps the most interesting—question that Durkheim failed to ask is what is hiding behind the fear of menstrual blood, which takes so many forms across our planet. Durkheim never asked or tried to answer this question because he viewed menstruation as one, private, instance of blood. But if this were so, he could just as well have related only to women's blood and left menstruation out of his theory. Evidently, menstruation has a special importance in this context which he could not reveal. And he is in good company. Many anthropologists and psychoanalysts do not know how to deal with it either. Thus, while the literature is replete with descriptions of menstrual taboos and rituals from all over the world, few researchers have tried to penetrate beneath the surface of the available facts.

The above notwithstanding, Durkheim's theory still appeals to me. I think that his connection between menstruation and the incest taboo is intuitively correct. In this connection, it should be mentioned that Robert Briffault has pointed out that all taboos evolved out of the menstrual taboo, which women forced on themselves and on all men.[18]

We have already discussed the deeper sources of the fear of menstruation and concluded that blood *per se* cannot be isolated from menstruation as a source of fears. Some anthropologists claim that the fear of blood is a myth. Briffault, for instance, states that not only is there no evidence for the fear of blood, but that blood is universally regarded as a delicacy.[19] While this is true of male blood, we have seen that female blood that comes from the genitals, even when it is not menstrual blood, is connected to the most frightening fantasies.

< THE LACUNA OF IMAGES >

In investigating how the incest taboo may be connected with the menstrual taboo, we should start by looking at male fantasies connected with defloration and sexual intercourse. Devereux's anthropological studies of the Mohave Indians tell us that there is a connection between fantasies about war and sexual intercourse. The Mohave male rituals after defloration are reminiscent of the cleaning of a sword after battle. And clinical material indicates that the male fear of destroying the female is sometimes a sadistic wish connected with the traumatic birth-destruction of the mother from inside.[20]

The fantasy of a child having intercourse with the mother from inside toward the outside and tearing her up is a reversal fantasy. It may arise when a mother tells a boy who fears being rejected by her because he is a male that he gave her a difficult time at his birth—something like 'you nearly killed me'. This fear, of being the cause of the destruction of the mother can lead to severe sexual disturbances.

Curiously, we have no analytic evidence regarding male fear of menstrual blood. (Klein, for instance, did not comment on boys' reaction to menstruation.) This is a missing link. But the fact that the menstruation taboo was originally imposed by men makes it obvious that men do have deep fears of menstrual blood. We may assume that they are basically the same fears as those of women, except that men feel that they are at the mercy of women for life or death. In this spirit, Karen Horney asks, "Does the man feel, side by side with his desire to conquer, a secret longing for extinction in the act of reunion with the woman (mother)? Is it perhaps this longing which underlines the 'death instinct'?"[21]

Sexual intercourse and the process of reproduction are associated in one way or another with blood, destruction and murderous wishes. The sexual and matrimonial bondage between men and women symbolizes in a way the disturbance of harmonious unity introduced by blood, pregnancy, birth, abortions and gynaecological complications. It is therefore plausible that incest and endogamy taboos relate at least partially to the unconscious fantasy that pregnancy and birth are associated with loss and death as much as they are with life. Blood is the common denominator of all this, and it is exclusively woman's blood. The embryo is in her keeping for life or for death. She is the source of both. The basic fears of humankind are connected with the inside, the place where life begins and from where it is eventually destroyed. This fear is expressed in the menstruation–pregnancy–parturition triad, which exists in many different societies. The menstruation and incest taboos are connected by the common denominator of destruction of the body from inside and the infanticidal element.

How, then, can harmony be preserved? Here is where Durkheim's

< INCEST AND MENSTRUATION TABOOS >

idea enters the picture. In order to ensure that both individual and clan remain intact, loss of blood must be prevented. This can be accomplished through a series of magical rituals aimed at preventing destruction—the menstrual taboo. But the menstrual taboo solves the problem only partially. Destruction could only be prevented by refraining from the act of sex, which is bloody in itself—in reality or in fantasy. The taboo against contact with menstrual blood, together with that against sexual relations or matrimonial relations within the clan, can prevent destruction.

In Durkheim's model, there is no separation between the individual and the collective. When the individual's blood is lost, it threatens the blood of the entire clan—the soul of the collective. The loss of blood must therefore be prevented by any and all means.[22] However, as menstruation cannot be prevented, the strict rituals of the taboo are necessary. The clash of blood through sexual intercourse can be prevented by the incest taboo and the law of exogamy. Both these taboos together solve the problem.

Human beings seem to have a deeply ambivalent attitude toward incest and endogamy. The urge to unite and fuse with the familiar, the need to preserve the symbiosis and fear of the 'other' all support the tendency toward incest and endogamy. Nonetheless, there is an almost universal repulsion or fear of these phenomena. These fears are both active and passive, directed toward the self and the 'other' at the same time. The fantasies of murder and self-annihilation must be shifted to the other, to another clan. Thus, sexual relations and marriage must take place outside the family, outside the clan.

The classical definitions of taboo refer to a condensation of the holy and the profane, both of which are dangerous. Taboos do not refer only to obedience to a law or even only to internalized prohibitions on the superego level. They express—and even sometimes constitute—the self.[23] However, if we accept that the incest taboos are basic to individual and collective existence, how are we to explain the weakening of these taboos?

There seems to be a developmental line from the tribal community to modern society, which is expressed in terms of a shift from the total identity between the individual and the collective toward a growing individuation. As the taboos cease to express the collective self, they become no more than internalized norms. And as individuation within the group increases, the meaning of the group changes and more exceptions to the rule are permitted. Individuals no longer fear that they will bring destruction to the collective by trespassing taboos. Adhering to the taboos is merely a matter for the superego and not a question of life or death in the most basic sense. It is my contention that the theory

< The Lacuna of Images >

connecting the incest and the menstruation taboos is the one that best explains the almost universal fear of feminine blood. In the psychoanalytic context, the fear of the feminine blood is fear of existence itself.

Not only does this hypothesis touch on the basic level of human motivations and sensations, but it succeeds in connecting two taboos in one theory, and the more phenomena a thesis succeeds in combining and explaining, the more significance it gains.

All the anthropological and psychoanalytic theories to date regarding this subject have shortcomings. For example, one may claim that Westermark's socio-biological theory disregards the basic observation that there is a sexual affinity between fathers and daughters, mothers and sons, sisters and brothers. A biological mechanism that may prevent incest in the animal kingdom does not seem to exist in humans. The statistics of incest in the twentieth century speak for themselves. Freud attacked this theory on the grounds that, if there were no potential for seduction, there would be no need for such strict taboos. Social prohibitions exist precisely to prevent the doing of things one wants to do.[24]

The alliance school can be faulted on the grounds that Lévi-Strauss's suggestion that the incest taboo functions in the service of economic interests (women for goods) relies on too rational a motive. After all, it is clear that the taboos in question work in the sphere of the deepest emotions.

Freud's family socialization theory, as presented in *Totem and Taboo*, is that the pact against incest stems from the guilt of the brothers in regard to patricide. The brothers of the primordial clan killed their father and ate him. After they satisfied their hate feelings toward him and fulfilled their wish for identification with him, they satisfied their feelings of love for him by eating him. According to Freud, their feelings of remorse led them to declare that the killing of the father substitute (the totem) was a transgression and gave up its fruits by forbidding themselves the women who came free. The guilt of the sons led to the two basic taboos of totemism, which parallel the two repressed wishes of the Oedipus complex—to preserve the totem and to refrain from marrying women within the clan.[25] Thus, Freud's theory is that the incest taboo and law of exogamy is based on guilt. But we know that guilt has never been enough to prevent transgression. The incest taboo demands a much more powerful theory, one that touches on the very existence of the self. For only such a theory suffices to explain the strength of these taboos and the basic fears that lie behind them.

The hypothesis I have represented obviously also has shortcomings. It does not explain what happens in societies in which there are no incest or menstruation taboos. Nor does it explain sufficiently why some

< INCEST AND MENSTRUATION TABOOS >

societies prohibit endogamy but have no incest taboo. Nonetheless, it seems the closest in spirit to human nature as seen from the psychoanalytic point of view.

I have presented Durkheim's theory of the incest taboo and law of exogamy, and my own thesis that there is a connection between the incest and menstruation taboos, to counter Freud's incest theory, which focuses entirely on males and completely neglects the female point of view. Likewise, I mean to challenge his theory about the menstruation taboo, which concentrates exclusively on the feminine aspect. Nobody since Freud has presented any theory that bridges this separation, due to the lacuna of the inside of the body. The fear of the woman, and the basic fears common to both sexes that are awakened by the female's productions, has, in turn, prevented any attempt to broaden discussion on the inside of the body that would give us more insight into these taboos, taboos that result from fear basic to human culture.

While many have expressed dissatisfaction with the existing theory, there has been no breakthrough because nobody has been prepared to violate the analyst-patient pact of silence with regard to the inside of the body. This is a taboo that psychoanalysis has inflicted upon itself. In this connection, Jungian analyst Erich Neumann remarks that the symbolism of the "terrible mother" draws its images predominantly from the inside.[26] And this terrible mother must be denied at all costs.

< THE LACUNA OF IMAGES >

PART III

Scientific Myths and Lacunae: From Thanatos *to* Logos

I propose to leave the dark and dismal subject of the traumatic neurosis.
— S. Freud

The chapters in Part III demonstrate the use of scientific myths in late Freudian theory, along with the connection between them and the existence of lacuna. The three building blocks of this mythology are: Eros, Thanatos, and Logos. Chapter 6 explains how Thanatos is born, what the motives were for the creation of the Thanatos theory, and which purposes it served. Chapter 7, describes the problematic nature of the dualistic mythic combination of Eros/Thanatos, and the use of the myth in order to strengthen the standing of Logos, which had been undermined by the new theory of drives.

6

The Deductive Birth of Thanatos *Theory*

The argument presented below exposes the manner in which psycho-analysis has consistently ignored the difficulty of explaining nightmares and recurrent post-traumatic dreams in the context of the Freudian theory of drives. This disregard, which may be considered a lacuna, has interfered with scientific progress. It gave rise to a chain reaction of diffi-culties in the development of the understanding of the dream mechanism, emotional trauma, traumatic neurosis and anxiety. A single scientific exception—the nightmare—created significant develop-mental scientific retardation in the discipline of psychoanalysis.

The function of Thanatos in the context of the new theory of drives is necessarily brought about, says Freud, by a variety of clinical reasons, which he details in his essay *Beyond the Pleasure Principle*. As I will show, the function of Thanatos is mythological, and it is, in fact, a scientific myth. Its unconscious goal is to solve *deux ex machina* the confusion relating to doubts concerning the integrity of the theory of drives. The lacuna around the post-traumatic nightmare brought about the creation of a scientific myth whose purpose was to preserve a theory that was about to be overturned because it was unable to explain a scientific phenomenon that was considered an exception.

Until 1920, Freudian psychoanalytic theory was based on a single theory of drives, the libido, which is the engine of emotional life and from which all its activity stems. This monistic theory is in turn based on the pleasure principle. The year 1920 marks a sharp turning point in Freudian theory. From that time on it is based on two drives: the drive to life—Eros, which in the past had been described as the libido; and the drive to death—Thanatos, which was added to it.

Freud was a clinician and believed that his theories were derived from clinical observation and should above all serve clinical goals. But do the changes in the theory of drives in fact derive from clinical motives?

This chapter will demonstrate the mythological growth of

< THE DEDUCTIVE BIRTH OF *THANATOS* THEORY >

Thanatos—the death drive—into the theory of drives, in the context of which I shall demonstrate the way in which mythological assumptions impose themselves on instinct theory and its function.

Freud asked himself about the nature of the repetition compulsion, which he thought more primordial and instinctual than the pleasure principle it downgraded.[1] He connected the repetition compulsion to the biological repetition of anabolism/catabolism. But was it this biological repetition that caused him to change instinct theory. If Freud had long been aware of this idea, why did this not happen until 1920?

Did something happen in the development of drive theory (based on the pleasure principle), something that forced Freud to shape the Thanatos theory and connect its drives to biological return? Beyond the biological perspective, there were several factors within the mental sphere that brought about the change in his theory of instinct: children's games; resistance to psychoanalytic theory and negative therapeutic reaction; the malignant fate of normal people; sadomasochism; the post-traumatic neurosis and dream. I shall examine each of these elements and analyse the extent to which each of them contributed to the need for a change in theory.

Children's Games

Freud asked himself whether, when children repeat a game in the effort to overcome a traumatic state, this repetition deviates from the pleasure principle? Does such activity fall within the framework of pleasure, or should it be considered an attempt to overcome a trauma? Or perhaps it serves both functions. In attempting to discern which of these factors has the most to do with a drive as basic as Thanatos, Freud struggled with himself. He doubted whether the drive to work out major psychic events and master them might turn out to be fundamental and therefore independent of the pleasure principle. But when he asked himself whether the instinct is connected to the compulsion to repeat, he determined that there is a fundamental tendency of the instinct to return a situation to its original state. This is an expression of inertia, a theme we shall meet later in the myth of Aristophanes. I ask myself why we shouldn't see in the wish for inertia an expression of Eros instead of Thanatos? After all, examples of pleasurable and archaic inertia abound. Repetition in play as well as in many other areas often expresses primary pleasure and not the auto-therapeutic attempt to undo a trauma. Moreover, the repetition compulsion can be demonstrated in every neurosis and transference neurosis.

Since all this was undoubtedly known to Freud, there is no reason to

< SCIENTIFIC MYTHS AND LACUNAE: FROM *THANATOS* TO *LOGOS* >

base Thanatos on the grounds of a universal repetition compulsion. Thus, children's games could not have provided a serious enough reason for Freud to reverse his instinct theory.

Resistance to Treatment and Negative Therapeutic Reactions

Is there anything in the resistance to treatment or in negative therapeutic reactions that makes Thanatos indispensable? Freud described situations in which patients forced him to treat them with coldness and rigidity, because they either wished to finish their analysis before it was time or placed obstacles on the therapeutic path. In conjecturing why patients flunk therapy, Freud came to the conclusion that it is the repetition of a painful act, not a pleasurable factor. Such repetition must therefore be viewed as destructive, and not as a way of trying to relive pleasure in a tacit way.

We must ask, however, if this is the only way to explain the resistance phenomenon. Freud, for example, did not ask himself why he (Freud) became cold and stiff when patients began to show signs of obstinacy. Maybe the dog is buried here. And what about the patient's fear of change? Perhaps the known evil is preferable to the unknown good? Why not explain this sort of conservatism on the basis of the pleasure principle? Resistance to therapy and negative therapeutic reaction were not new to Freud. Why did he now decide to explain them with Thanatos when he had lived with them as part of the pleasure principle for so long?

The Malignant Fate of Normal People

And what about the malignant fate of normal people[2] possessed by some demonic power. Freud takes up the example of benefactors who are abandoned in the end by their friends. What analytic explanation can there be for such a phenomenon? Can normal people who suffer from a repetition compulsion be described as having 'demonic traits'? Freud's wording, which resorts to a term that lies in the realm of the inconceivable, the mythic or the primordial, would imply that there is no psychoanalytic explanation for the repetition compulsion. But this is precisely Freud's tacit thesis: that whatever cannot be explained analytically is mythical and falls within the domain of Thanatos. Is Freud implying that being left by one's best friend cannot be explained analytically? Is he referring to himself? Perhaps people who experience this phenomenon repeatedly should ask themselves how they choose their

< THE DEDUCTIVE BIRTH OF *THANATOS* THEORY >

friends—among other questions. Must we presume that this experience lies within the domain of Thanatos and outside the region of Eros?

Sadomasochism

In *Instincts and Their Vicissitudes*,[3] Freud introduced sadism as a primary instinct and masochism as a secondary instinct within a double framework. He considered masochism as the authentic mechanism of sadism—all within the framework of the trusty old instinct theory. Since nothing new occurred in the development of the clinical theory that would oblige Freud to change course by reversing this theory, it is doubtful whether the question of primacy is amenable to any solution on clinical grounds.

As we have seen, the silent aspect of Thanatos has not so far explained anything new. Indeed, at first sight it seems to express much more the psychoanalytic silence when faced with phenomena for which there is not yet any explanation, or phenomena which have aroused resistance.[4]

Freud did not generally yield to blurred situations or accept demonic traits as a given. After all, he spent his life rooting them out and fighting against them. They are part of the type of mysticism he wanted to save us from. Nonetheless, it would appear that, although Thanatos does not hold first place in psychoanalytic theory, it has been incorporated into the theory as an axiom whose mythical essence is exempt from the need of proof. It could therefore be chosen arbitrarily. Judging by the examples given so far, Thanatos does not seem indispensable to the theory, but somehow Freud was not satisfied to leave it within instinct theory.[5]

Post-Traumatic Neurosis and Dream

Let us turn now to what seems to me the real reason for Freud's shift in the drive theory—the post-traumatic neurosis and dream.

The most serious problem we encounter in instinct theory is the fear that the repetitive nightmare can no longer be explained on the basis of wish-fulfilment. For nightmares did not seem to fit into the framework of Freud's dream theory. Freud endeavoured to circumvent this embarrassing problem by claiming that repetitive nightmares differ from regular dreams and do not therefore contradict dream theory.

Not being able to discern the wish element in post-traumatic dreams, Freud determined that they are exceptional dreams. He evolved two different theories to explain their function. The first, mastery theory, suggested that post-traumatic dreams are a repetitive working through

< SCIENTIFIC MYTHS AND LACUNAE: FROM THANATOS TO LOGOS >

of the trauma. Alternatively, he viewed such dreams as indicative of a temporary defect in the dream mechanism that does not allow it to function as wish-fulfiller. Freud claimed the second hypothesis did not contradict his monistic dream theory because it is temporary and the usual dream function would eventually return. In support of his first theory, he claimed that post-traumatic dreams revive the signal anxiety that was damaged by the traumatization. Because traumatic neurosis produces traumatic anxiety, while the dream produces signal anxiety, the latter functions in the service of the ego—and therefore within the realm of drive theory.[6] However, this notion contradicts clinical observations on traumatic dreams which should have been known to Freud: the anxiety is not functional. We shall return to this point.

Freud eventually discarded his mastery theory in favour of the defect theory and no longer mentioned it in his writing. He was forced to refute this theory when it became clear that it did, indeed, contradict monistic dream theory, not because of new clinical observations. Later, when he returned to the problem of wish fulfilment, he proposed that post-traumatic dreams are attempts at wish fulfilment. Freud went to great lengths to preserve his monistic dream theory. When he was unable to bridge the gap between post-traumatic dreams and ordinary dreams to his satisfaction, he "proposed to leave the dark and dismal subject of the traumatic dream".[7] However, by leaving the traumatic neurosis out of his theory, it remained, like a ghost, in the box.

How did Freud's disciples react to this embarrassing situation? Anyone summing up the psychoanalytic literature on nightmares, especially on those of the post-traumatic type, must conclude that the subject has not received the clinical and theoretical attention that a phenomenon of its importance would seem to demand. Psychoanalysis did not realize that post-traumatic dream and nightmare in general constitute a veritable gold mine when it comes to exploring the function of dreams, traumatic neurosis, anxiety and the nature of repetition compulsion. We learn from the history of science that exceptional scientific problems that do not fit into the paradigm generally unleashed a curiosity to conquer the unknown, and that this often stimulates new processes of discovery.[8] However, with respect to the post-traumatic dream in psychoanalytic theory, the opposite seems to be the case. Instead of examining these phenomena precisely because it represented a challenge, Freud and his disciples avoided dealing with it by sweeping it under the rug—in contradiction to usual developments in science as well as in psychoanalysis itself.

The problems that arise in the study of post-traumatic dreams lead to areas outside the traumatic dream itself. In the dream, we observe the clinical uniqueness of the anxiety aroused both during the dream and

< THE DEDUCTIVE BIRTH OF THANATOS THEORY >

upon awakening. We also observe the undisguised or seemingly undisguised nature of the manifest dream and the compulsive, rigid repetitiveness of the traumatic situation. At least at first glance, the traumatic dream does not seem to follow the paradigm of the dream as put forth in dream theory. For the regulatory mechanism of sleep conservation fails to function here, while, according to classical theory, the dream functions as sleep guardian, and the resultant anxiety does not seem to be under the control of the autonomic nervous system. An understanding of these dreams might bring about a better understanding of the basic difference between signal and traumatic anxiety.

The questions that might have been asked about the nature of the post-traumatic dream spill over into the area of dream theory in general and extend to other areas of psychoanalytic interest. One might begin by inquiring into the function of the traumatic dream in general, because it raises a question about wish-fulfilment theory as it relates to dreams. Then one could ask whether these dreams are always symbolic in nature, or whether they may not sometimes be a-symbolic. For nightmares, especially of the post-traumatic type, may show us much that could open up the way to a better understanding of phenomena in several other areas; for example, a great deal could also be learned from traumatic dreams about the nature of the repetition compulsion and the significance of the manifest content of the dream, a subject discussed in the literature, but not from the point of view of the traumatic dream.

Some of Freud's followers did endeavour to provide an alternative to his explanation of the essence of the traumatic dream. Thus, Max Schur concluded that traumatic dreams represent the ego's attempt to undo the trauma,[9] and Martin Stein suggests that the function of such dreams is to deny it.[10] According to Stein, the repetitive dreaming allows the dreamer to feel upon awakening that the traumatic experience was 'just a dream' and thereby deny the existence of the trauma. From this we may conclude that, under certain circumstances, the function of the dream is to awaken rather than to guard sleep. Theodor Lidz argues that the traumatic dream may express wish-fulfilment, if we hypothesize that the wish to die, which often appears in the dreams of post-traumatic soldiers, expresses a vital wish, which is not as paradoxical as it might sound, for under battle conditions, with its stress and the death of comrades, death is taken for life and life for death. Under such circumstances Lidz does not view the wish to die as something beyond the pleasure principle, but as a wish among other wishes.[11] In so doing he negates Thanatos theory.

We see that not only Lidz, but also Schur and Stein, speak openly of the wish in these dreams. They do so in an attempt to get Freudian theory back on track by showing that, in the end, the traumatic dream

< SCIENTIFIC MYTHS AND LACUNAE: FROM *THANATOS* TO *LOGOS* >

is a wish-fulfilment dream ruled by the pleasure principle. Denying that the trauma experienced has been undone expresses the wish that the terror never really existed. John Mack gives a different and perhaps more comprehensive explanation of the essence of the traumatic dream when he points out that the mechanism involved is one of turning against the self. In his view, the aim of the dream is to turn the aggression aroused by the traumatic experience back against the self in order to keep it from bursting out against the outside world in fantasy or reality, which might become too threatening for the ego.[12] Mack's theory can be included within the framework of the pleasure principle, as the dream expresses a defence against aggressive drives and therefore fits what Freud characterized as masochistic dreams in the service of that principle.

In general, these explanations all attempt to conceptualize the traumatic dream within the framework of the pleasure principle, in order to salvage Freud's wish-fulfilment theory. However, one doubts that any of them can, on their own, convey the whole essence of these dreams.

The suggestion that denial and undoing mechanisms are the only dream functions raises the question whether dreams that reproduce traumatic experiences and anxieties can really be viewed as the legitimate products of these mechanisms. The experience during the traumatic dream and upon awakening produces anxiety of such a nature that Mack feels that such dreams may sometimes be considered psychotic.[13] In this context, the tendency of post-traumatic soldiers to avoid confronting their dreams by any means possible is itself an indication of the importance of their impact. It might be argued that these dream experiences are a defence against the more severe experience in reality, and that if the dreamers were deprived of their dreams they might become psychotic. However, even if this were the case, it is difficult to see what roles denying and undoing play in this process. Moreover, it should not be taken for granted that the dream is not a traumatizing experience in itself.

Regular dreams, including anxiety dreams, are considered to play a major role in maintaining the integrative functions of the ego. Is this also the case for post-traumatic dreams? Nightmares and night-terrors are usually viewed as the actualization of man's deepest fears. The signal anxiety apparatus fails. Upon awaking, the dreamer remains in a state of terror, experiencing extreme tachycardia, transpiration and disorientation in time and place. Sometimes hallucinations and apnea occur. The dreamer develops a sleep phobia, avoiding sleep in order to avoid dreaming. The extreme tachycardia (sometimes more than 200 beats a minute) can lead to cardiac complications and sometimes even to death. Nightmares and night-terrors are often looked upon as temporary

< THE DEDUCTIVE BIRTH OF *THANATOS* THEORY >

psychotic states.[14] All these facts indicate that post-traumatic dreams are not controlled by the vegetative system. They lack the necessary home-ostasis, which, in ordinary dreams, enabled the dream to carry out its functions. This would mean that post-traumatic dreams are basically different in any sense from normal dreams and could not possibly have the same function. The most elegant worker in this field, Charles Fisher, has clearly demonstrated that the REM functions as a central regulator of the vegetative system, which explains why, biologically, the anxiety in anxiety dreams never goes beyond a certain limit, for, whenever this threatens to happen, the alarm system awakens the dreamer. This discovery, together with his findings that NREM dreams do not come under the control of the central nervous system as the signal mechanism does not work, indicates that there are ways to define post-traumatic dreams and neuroses in a new light.

Physiologically, traumatization threatens to erupt whenever the vegetative system escapes central REM control. When this occurs the organism is unable to activate the desomatization mechanism that keeps anxiety within a bearable range, that is, one which would not cause harm to the biological systems. To put it in terms of adaptive versus non-adaptive anxiety, the inability of the buffer system to work may be viewed as an analogue of the ego's failure to regulate the different dream-mechanisms, including the anxiety signal.

Fisher has also shown that nightmares themselves are symptoms which may then produce a traumatizing effect. This would accord with the phobic reaction of post-traumatics toward sleep.[15] However, the mechanism of the post-traumatic dream itself has barely been explored. At this point we are limited to considering them on the basis of our knowledge of nightmares and night-terrors, and on clinical observa-tions. What we can say, nonetheless, is that some dreams do have traumatizing effects, and that this is in opposition to classical dream theory.

While Mack's contention that the dominant mechanism in post-trau-matic dreams is the turning against the self may explain war and post-accident neuroses in which a clear element of guilt can be traced, this is not always the case. When the superego is not well developed and guilt is not the main theme, or where aggression does not play a main conflictual role, such as in narcissistic or dependent personalities who suffer from the same repetitive traumatic dreams following trauma, turning against the self cannot be regarded as a defence characteristic. Since we may argue that elements other than guilt can play a role even when the superego is well-developed, I doubt that we can accept Mack's theory as a general one.

The main criticism of the mastery theory as I see it is that the theory

< SCIENTIFIC MYTHS AND LACUNAE: FROM *THANATOS* TO *LOGOS* >

views mastery as the sole determining function despite the fact that post-traumatic dreams have been known to continue for as long as thirty years (in the case of those suffering from post concentration-camp syndrome). Thus, when speaking of attempted mastery or the failure of attempted mastery, we must not fail to identify the contradictory force in the dream that prevents mastery. This force may perhaps be identified with service against the self (instead of being in the service of the ego).

The Asymbolic Dream and Repetition Compulsion

The discussion of post-traumatic nightmares raises the question whether asymbolic dreams exist. Asymbolic dreams stand in apparent contradiction to the basic psychoanalytic concept of the very essence of the dream. This may be why the asymbolic dream is granted a kind of second-class citizenship in psychoanalysis. Freud directed our attention to asymbolia when he dealt with aphasia,[16] identifying it as present in second-order aphasia, wherein the association between word and object presentation is disturbed. In such a case, asymbolic is correlated with object-relation disturbances, while symbolism is connected to communication.

Freud's writings indicate three categories of asymbolic dreams, two of them in *The Interpretation of Dreams*.[17] The first of these relates to the pre-verbal phase, when the child is still not capable of symbolizing. The second is the dream of convenience, a dream that may be dreamt in order to maintain sleep (always its function, according to him) without necessarily being symbolic.[18] However, his recognition of such asymbolic dreams was *de facto*, since he never referred to them as such. The wish to isolate dreams of convenience from symbolic wish-dreams is a wish in itself. Nor did Freud make any effort to identify the symbolic function in these dreams, fearing that he would not be able to find it, or, perhaps he thought that hallucinations to preserve sleep are a special category. If this was the case, he violated monistic dream theory. Freud never made it clear where we should draw the line between so-called or not so-called asymbolic dreams and regular dreams. And why should we not consider it a violation of general dream theory to make this distinction?

The third category of asymbolic dreams may be identified in some post-traumatic dreams, although neither Freud nor his followers identified them as such. Their possible existence is derived from such descriptions as the following: "Shock dreams usually do not undergo any distortion or any other elaboration, which leads us to ask whether

< THE DEDUCTIVE BIRTH OF *THANATOS* THEORY >

the ego has any active share."[19] These dreams, which every clinician that works with post-traumatic patients may experience, are realistic, bright and repetitive in a stereotyped way. The special meaning of this sort of repetition compulsion is discussed by several writers, who never ask what implications this may have for dream theory. Freud's texts do not distinguish between post-traumatic repetition and other neurotic repetitions, such as repetition in transference or in everyday life, such as children's play. The difference is so sharp that it is strange that Freud did not delineate it.

Freud never referred to the possible asymbolic nature of the post-traumatic dream, although it can be deduced from his theory of temporary dysfunction. Making such a statement would bring into focus and sharpen the conflict with general dream theory. In 1966, Laurence Kubie stated that, both awake and asleep, there is a constant imageless and asymbolic, preconscious stream of central activity, consisting of a continuous subliminal processing of experience.

As early as 1939, Kubie labeled the term 'repetition compulsion' unclear and claimed that it was an 'uncontrolled way'. He charged: "Everything which occurs in mental life is very likely to be dropped into this scrap basket!"[20] Since he believes that the post-traumatic dream is an attempt to take hold of the traumatic situation, he cannot see any difference in the way the repetition compulsion works in such dreams. In his view, the post-traumatic neurosis is just another neurosis. Even earlier, Leon Saul did not see any qualitative difference between the two repetition compulsions because there is always a symbolic element in repetitive dreams, as it never reflects exactly the real situation.[21] One can say that the tiny difference from one traumatic dream to the other can be considered as a symbol. This symbolic element is the sign of ongoing primary process activity that might open a door to interpretations.

On the other hand, Eduard Bibring distinguishes clearly between repetitive dreams with a restitutive element and those with a fixative element. According to him the repetition compulsion can be seen as an instinctual drive of the id, the ego, or of both. Related to the id, it would involve an instinctual automatism that we would refer to as fixation or tension repetition. Related to the ego, it would be of the regulatory or restitutive type.[22] Robert Waelder described the repetition compulsion in a descriptive manner by calling it Janus-faced.[23] Bibring disagrees with Alexander insofar as the post-traumatic dream and revisions in the primary processes are concerned. In his view the traumatic dream is a clear and accurate reproduction of the traumatic experience, while Alexander believes that the situations described in battle dreams are worse than the reality.[24] Thus, although Bibring claims that post-traumatic dreams are asymbolic and without elaboration, he does not

< SCIENTIFIC MYTHS AND LACUNAE: FROM *THANATOS* TO *LOGOS* >

develop this idea *vis-à-vis* general dream theory. It would appear that this would violate a taboo.[25] Any reconsideration of Freud's dream theory is impossible.

One cannot understand the essence of the repetition compulsion in the traumatic neurosis without identifying the fixation repetition in it. The asymbolic state gives the repetition compulsion a different meaning than regression in the service of the pleasure principle. It is in fact actually a regression from life, a disconnection from communication and therefore from object-relations. The psychic apparatus that usually functions as a means for elaboration of psychic experiences ceases to function as such. When there is no search for new ways, there is a regression to the catabolic state. And whether or not the death instinct exists, catabolism of the psyche to the traumatic state gives Freud's 'demonic idea' a very specific and accurate meaning.

In the debate between Franz Alexander and Eduard Bibring, truth seems to be on both sides. The post-traumatic dreams show in many cases a development over time, but this is not always the case. When there is no development we do not succeed in getting free associations about the dreams simply because they are probably asymbolic. The fact that these dreams often appear as films, reflecting very clearly the same realistic situation over and over again, suggests that the dream apparatus has 'broken down'. To my knowledge there are no clinical experiments verifying this idea, only clinical impressions. Nonetheless, Ernest Hartman's findings, which suggest that the post-traumatic dreams of patients who showed the clearest and most severe signs of traumatic neurosis reflected the memory of the trauma most accurately, indicate that these dreams have not undergone any sort of working through. Hartman found that the nightmares of patients who improved over time gradually changed from representing direct memory to a more distorted, symbolic, 'dream-like' reference to the traumatic event. Thus, clinical improvement seems to go hand in hand with the capacity to symbolize.[26]

It is paradoxical that traumatic dreams are at their most naked and manifestly seem to be unable to solve the riddle, while dreams that are more disguised and confused are much more helpful. But this paradox is exactly what Freudian theory is about.

The possible asymbolic state of post-traumatic neuroses can be deduced not only from the dreams produced. For instance, if we observe the startle reaction[27] so common in traumatic neuroses, we see that they might be evoked in consequence of only minimal stimuli. This reflex is reminiscent of the baby's Moro reflex—a massive, total, unregulated body reaction to noise. Once the stimulus breaks through the stimulus barrier, the baby is overwhelmed and loses control. In its worst form the

< THE DEDUCTIVE BIRTH OF *THANATOS* THEORY >

startle reaction indicates that the vegetative system works in a primitive, vehement and uncontrolled way, signalling an unmasked and traumatic state. The post-traumatic neurosis and dream reflects this by exhibiting asymbolic, uncontrolled features, beyond the pleasure principle, and by expressing the most archaic psychophysiological existence.

The asymbolic traumatic state receives support from two analogous situations: The first situation involves reaction to treatment with penthotal. Under the influence of this psychopharmacological drug, patients re-enact traumatic situations as they endured them. They return to the same situation and to the same affective reactions—without elaboration.[28] The second analogous situation concerns pseudo-hallucinations. Pathognomonic to the traumatic neurosis, pseudo-hallucinations often reflect the traumatic situation in the form of an internal stimulus. They are described by the patients as a very vivid copy of the original situation.

In fact, there are three analogous phenomena that point to an asymbolic state: (1) the post-traumatic dream; (2) reaction under penthotal; and (3) pseudo-hallucinations.[29]

If the failure of the post-traumatic dream does not always necessarily derive from the conflict that caused the traumatic reality, to what shall we attribute this failure? It may result from a defect in the dream mechanism, whether primarily due to a psychic factor, a biological factor or a combination of both. This defect manifests itself in the incapacity to work out intra-psychic experiences by means of primary process mechanisms like symbol formation. It may be a temporary defect in the synthetic capacity of the dream, in which the coming together of the traumatizing factor and the psyche produce indications that cannot be worked out. This results in a broken record, a stereotyped repetition of the traumatic reality. Where this is the case, there may be a primary process thinking defect, which prevents elaboration, and working through—and this may explain the impossibility of producing transference effects or any sort of working through at this stage. As long as the situation continues, psychoanalytically oriented psychotherapy simply does not work.

Such ideas find little acceptance in the psychoanalytic community because they do not fit the assumption that the primary processes are always there and at our disposal. The notion of the asymbolic dream is particularly disturbing to those who view dream theory as based exclusively on drive theory and wish-fulfilment. Ego psychology did not correct this situation.

The only serious attempt to unify biological data concerning post-traumatic dreams with psychoanalytic dream theory was made by Max

< SCIENTIFIC MYTHS AND LACUNAE: FROM *THANATOS* TO *LOGOS* >

Stern in the 1980s. While it is not within the scope of this work to evaluate Stern's effort, I shall mention some of his theses.

According to Stern, the panic, dread and paralysis of the *pavor nocturnus* resemble the catatonic state of shock, the origins of which are in the central nervous system. *Pavor nocturnus* is a defence against stress when the immature ego is incapable of dealing with severe conflict (trauma) without arresting its development. "Traumatic dreams are not produced by psychic conflicts, rather they can be traced back to arrests in psychological development resulting from a lack of co-ordination with an original gratifying reality."[30] This is a common conclusion for self-psychology and Stern's research, as Levin remarks in his introduction to Stern's book.

Stern's claim is that primary process products do not reappear after the occurrence of trauma until the signal anxiety regains the mastery upon which repression is possible. The recovery of the primary process is dependent on the recovery of the secondary process. This brings us back to Freud's consideration of post-traumatic dreams as phenomena that are temporarily not under the sign of wish-fulfilment.

The fact remains that for dozens of years there was no serious attempt to take a fresh look at the exceptional position of post-traumatic nightmares, with or without the new biological research, until Stern was able to free himself from the burden of monistic dream theory and explain post-traumatic dreams independently from the wish-fulfilment.

Traumatic Neurosis

Traumatic neurosis is a stepchild in psychoanalysis. Although not a rare phenomenon, as it occurs quite often during wars and as a result of accidents and natural disasters, psychoanalysis has not evinced much interest in it. Indeed, Allan Compton claims that actual neuroses have received relatively little attention since Freud.[31] Thus, very few analysts have tried to conceptualize the phenomenon, or to compare it to other types of neuroses or psychopathological categories. As the general tendency was to treat it as an ordinary neurosis, few tried to find a connection between traumatic neurosis, the post-traumatic dream, trauma and anxiety.

Although Freud did identify the traumatic neurosis as unique, describing it as a narcissistic neurosis,[32] this description does not remove it from the bounds of drive theory. The drive seems to hide, but it is there—in the ego.

There is no doubt that Freud, as well as Karl Abraham and Sandor Ferenczi, realized that this neurosis is qualitatively different from other

< THE DEDUCTIVE BIRTH OF *THANATOS* THEORY >

neuroses, but they could not identify the difference. Why was this? There was no lack in those days of descriptions of traumatic neurosis. The First World War supplied a multitude of examples. Ferenczi even documented many observations from his own clinical experience.[33] It is no accident that the Budapest Congress of 1920 devoted an important place to the post-traumatic neurosis. In his opening lecture, Freud mentioned that those opposed to psychoanalysis were anxious to show that war neuroses did not fit drive theory:

> The term "sexuality" is to be taken here in the broader sense customary in psychoanalysis, and not to be confused with the narrower sense of "genitality". Now, it is quite correct . . . that this part of the theory has not hitherto been demonstrated in relation to the war neuroses . . . The opponents of psycho-analysis, whose repugnance to sexuality has shown itself to be stronger than their logic, have tended to proclaim that investigation of the war neurosis has finally disproved this part of the psycho-analytical theory . . . If up to the present superficial investigation of war neuroses has not shown that the sexual theory of the neuroses is correct, that is quite another matter from showing that this theory is incorrect.[34]

It seems that Freud's disciples were more Catholic than the Pope in dealing with the post-traumatic neurosis. After all, Freud did not conceal that it confronts us with a serious theoretical problem, and he continued to think about the problem and discuss it as such. But his disciples and the following generation blurred the problem. Thus, Ferenczi was adamant that it was a regular neurosis. After describing clinical observations in some detail, he concluded: "My object is achieved if I have succeeded in showing you that the clinical pictures presented to you do belong to those two disease groups that psycho-analysis has designated anxiety hysteria and conversion hysteria."[35]

Ernst Simmel, who also spoke about his clinical experience with war neuroses at the Budapest Congress, treated soldiers by interpreting their dreams, which he claimed were attempts at self-healing. He did not distinguish them from anxiety dreams.[36] Roy Grinker and John Spiegel, who had much clinical experience with traumatized soldiers during the Second World War, did not identify any pathognomonic qualities. They included the war neurosis in the psycho-neuroses.[37]

Abraham Kardiner was first to find qualitative differences between post-traumatic neuroses and psychoneuroses. In his view, the post-traumatic neurosis is a stereotype, and symptoms in addition to the post-traumatic dream itself are incapable of symbolic extension (displacement).[38] Still, he did not clear up the nature of the traumatic anxiety.

Freud and others considered post-traumatic neuroses as a form of

< SCIENTIFIC MYTHS AND LACUNAE: FROM *THANATOS* TO *LOGOS* >

regular neurosis in which the current function of the mental apparatus is suddenly disrupted due to external occurrences. His term 'actual neurosis' is confusing, for we can discern trigger mechanisms stemming from the external world in every neurosis. The question here is how the external world affects the intra-psychic apparatus in a particular case, at a specific time. After all, different soldiers exposed to exactly the same external stimulus may have totally different reactions, and, even when the stimulus is extremely severe, not all of them collapse into traumatic neurosis.

Freud's division between 'ordinary' and 'actual' neurosis only masks the problem. The very term 'actual neurosis' indicates psychoanalysis's lack of interest in phenomena that indicate there is not much to be found in the internal world of fantasy. The 'actual neurosis' brings us back to so-called reality, which has little appeal to a discipline that occupies itself with intra-psychic processes. Nonetheless, there is no escape from the fact that the dichotomy between actual neurosis and psychoneurosis is artificial, for what counts in the end are not the stimuli in themselves, but their intra-psychic consequences, whether they come from outside, inside or both. Reality is always translated into the inner world in an idiosyncratic way.

Anxiety and Trauma

Seventy-five years after the Budapest Congress psychoanalysis has not succeeded in crystallizing a theory of anxiety that sheds light on difficult clinical phenomena and theoretical problems. One of the central problems in the theory of anxiety is related to the meaning of the term 'trauma'. Another is the failure of psychoanalysis to evolve a clear concept of trauma and its reciprocal relationship with anxiety. I contend that the broad and blurred concept of trauma used by Freud's followers resulted from their failure to investigate sufficiently the interconnections between trauma and anxiety. This blocked the possibility of arriving at a psychoanalytic concept of the traumatic neurosis.

The definition of trauma extends at present from the original notion of the breakthrough of the stimulus barrier of one extreme to the notion of the accumulative, the strain, the retrospective, the screen trauma, until it becomes difficult to differentiate between adverse pathogenic influences in general and trauma in particular. This demonstrates the confusion that surrounded one of the central issues in psychoanalysis.

Investigating the concept of the individual *vis-à-vis* the term 'trauma', Joseph Sandler *et al.* found that it is still used to describe four different phenomena: (1) the overwhelming event; (2) the symptomatic results of

< THE DEDUCTIVE BIRTH OF *THANATOS* THEORY >

an external event (here the emphasis is on phenomenology); (3) a painful experience resulting from an external event; and (4) a diagnostic criterion as distinguished from psychoses and neuroses, that is, the post-traumatic neurosis.[39]

When we speak of trauma, do we mean a cause or a result? In the latter case, it is not clear whether we refer to objective clinical symptoms or to the subjective experience. Do we consider a psychic event a trauma solely when it originates from external stimuli, or when it originates in internal stimuli as well? Do we mean acute events or a chronic situation as well?[40] In its vague and diffuse use, the term lacks the qualities that might allow us to define specific causes or results. As such, in the present, it does not really explain any phenomena adequately.

A scientific term is significant and has value only if accurate enough to explain a phenomenon so that the phenomenon becomes distinct from others. Sandler *et al.* tried to solve the problem by finding a common denominator in the definitions and attitudes toward the term.[41] But their attempt bypassed the problem because it never reached the root of the phenomenon: the connection between trauma and anxiety.

In order to understand the true nature of the traumatic neurosis, we must first investigate this connection and then form a comprehensive anxiety theory. The questions to be asked are: (a) whether the two anxiety theories of Freud are mutually exclusive or are amenable to being united; (b) what the essential difference is between traumatic neuroses and regular neuroses, and how this can be explained in analytic terms; (c) whether the term 'trauma' can be defined more accurately and specifically? (d) whether there is a basic difference between traumatic anxiety in normal development and in the traumatic neurosis; (e) whether the post-traumatic dream, which is pathognomonic to the traumatic neurosis, can be taken as a model to help us better understand trauma.

In questioning whether the signal anxiety of the ego may be identical to the anxiety from which it tries to save the ego (the traumatic anxiety), we must ask whether transformation between these two states of anxieties is possible. Freud's general theory of neurosis teaches us that the neurotic symptom saves us from traumatic free-floating anxiety. If the free-floating anxiety of the neurosis were identical to the anxiety that occurs in traumatic neurosis, anxiety would be at one and the same time the pathogenic source and the factor that saves us from this source. In that case, we remain with the dilemma of whether we are dealing with one kind of anxiety or two.[42] In his *New Introductory Lectures* Freud claims that anxiety is a reproduction of an old event involving a feeling of danger. Not only does it serve the function of self-preservation, but

< SCIENTIFIC MYTHS AND LACUNAE: FROM *THANATOS* TO *LOGOS* >

it also works as a signal for a new anxiety. Anxiety results when the libido cannot find its place, and it is awakened during the development of repression and the emergence of symptoms. In the statement "Anxiety is on the one hand an expectation of a trauma and on the other hand a repetition of it in a mitigated form, " Freud describes anxiety as a Janus-faced united entity. But, earlier, he suggested that anxiety may have a twofold origin—one as a direct consequence of the traumatic moment, and the other as a signal threatening a repetition of that moment. Leo Rangell, however, points out that Freud never withdrew his earlier opinion nor solved the problem to his satisfaction. His followers merely overlooked the dilemma.[43]

In attempting to unite both theories, Rangel claimed that anxiety is always both signal anxiety—a warning from psychic trauma—as well as a reaction to its own presence that can vary from a minimal discharge by the ego, in a controlled and experimental way, to a moderate or large degree of already present helplessness. According to Rangell, anxiety is a traumatic reaction of helplessness, past and present, filtered through the perceiving, judging, reacting and anticipating ego, which adds its estimate of the future. Thus, he says, it always involves a traumatic situation at work simultaneously with controlling factors of the ego. Rangel's theory has never been clinically proven. Indeed, experience with patients suffering from post-traumatic dreams may indicate the contrary. Thus, Rangell's thesis does not solve the problem because it does not come to terms with the entity 'trauma' and the helplessness connected to it.[44]

In seeking a way to handle the unsolved riddle of the relationship between signal anxiety and traumatic anxiety, Clifford Yorke drew a developmental line of the condition of anxiety. Beginning with diffused somatic excitation as it occurs at the beginning of life, anxiety develops through pervasive psychic excitation till, with some ripening of the ego, signal anxiety is formed. Yorke distinguishes between three different categories of anxiety: (1) arrested anxiety which remains pervasive; (2) signal anxiety, which occurs when there is temporary or permanent regression to primitive stages of psychic helplessness and pervasive anxiety; and (3) still more regression to still more flooding with vegetative excitation. Where psychic functioning is minimal, he believes that traumatic anxiety is experienced by every child at every stage of development, but that it is rapidly reversible and therefore not a traumatic experience.[45]

From the above, it is clear that Yorke uses the term 'trauma' in two different ways. According to him, the child evinces traumatic anxiety, but it is not a traumatic experience! Nor is it clear how he distinguishes between the reversibility of a normal child's traumatic experience and

< THE DEDUCTIVE BIRTH OF *THANATOS* THEORY >

the irreversibility of 'real trauma'. (Here it should be stressed that many traumatic neuroses are reversible.) Yorke claims that in traumatic neurosis the ego is not merely flooded with pervasive anxiety, but totally 'knocked out' by a flood of excitation. What this 'knock-out' is, and how it can be defined in physiological and psychological terms, is not clear. We have no evidence that neonates cannot be totally flooded with overstimulation (as a precursor of anxiety).

I contend that this blurring of trauma and anxiety theory is not meaningless, and that the inconclusiveness concerning anxiety theory—indeed, the avoidance of trying to solve the problem—had a function, albeit an unconscious one. The question whether traumatization is a process that contradicts drive theory had to be left vague because the psychoanalytic community was following Freud's own inconclusiveness in this matter. Any unification of the two anxieties would necessitate a clear distinction between them, which would sharpen the conflict surrounding his general theory of neuroses.

Obstructions in the Development of Dream Theory

Let us return to the influence that the exceptional position of the post-traumatic dream had on the development of dream theory. If we look at the development of dream theory during the last ninety years, we might conclude that not very much new has occurred in this field. Although Freud revised this theory, as he revised his other theories, he also admitted that his revision brought little that was new.[46]

> In the introduction to his *The Revival of Interest in the Dream*, Robert Fliess writes: Freud's complaint of the lack of interest in the dream is well known. He accused us in the first part of his *New Introductory Lectures* of behaving as though we had nothing more to say about the dream and as though the whole subject of dream theory was finished and done with.[47]

This remark is still relevant today. Although much has been written on the place of the ego in dreams, on the dream as a means of communication, on dream theory in connection with psychophysiological research, on the manifest dream, and on the dream as a therapeutic tool, our understanding of dreams is not in essence much better now than it was in 1900. This is especially true of our understanding of the function of the dream and its mechanism. Despite ongoing developments in our understanding of the place and share of the ego in the personality structure, as well as our *de facto* recognition of the ego's rights in the dream, we have not yet accomplished a reformulation of the dream function that would be more useful from the clinical point of view—one that

< SCIENTIFIC MYTHS AND LACUNAE: FROM *THANATOS* TO *LOGOS* >

would take into account those stepchildren of the dream, the nightmare and the post-traumatic dream.

The traumatic dream represents only one example of the absurdity of viewing wish-fulfilment as the dream's sole function. In many cases wish-fulfilment does not even suffice to explain the material in ordinary dreams. Our recognition that a single dream can express id as well as ego forces, and the consensus that dream language represents the integrative functions of the ego, is based on analytical experience as well as on dream-deprivation research. It is therefore quite clear that a broader model is called for. The only real attempt at an alternative theory (besides self psychology) was made by Angel Garma, who argues against the validity of the wish-fulfilment theory, confronting it with the trauma theory.[48] But, here too we have a theory based on a monistic model.

A comprehensive and multifunctional model of dream functions seems to be obligatory for the following three reasons:

(1) As we have just seen, the study of the traumatic dream shows us that it cannot be satisfactorily explained on the basis of a single function. This type of dream can only be explained on the basis of a multifunctional model. But, as the traumatic dream cannot be separated from other dreams, there is no reason why we could not apply such a model to all dreams.

(2) Analytic theory is based on the meta-psychology of a multiple system. Pinchas Noy has demonstrated that the uniqueness of the psychoanalytic system, as opposed to other psychological methods, lies in its being based on a complementary multimodel system. He demonstrates the advantages that such a model brings to the natural sciences, for example, how it is the only model that satisfactorily explains the phenomenon of light as both a particle and a wave. He goes on to show how theories based on a single model are short-lived because they fail to explain the full spectrum of the phenomena involved.[49] But psychoanalysis, which is also based on a complementary multifunctional model, has not adopted this principle with regard to the phenomenon of the dream, instead choosing to follow a monistic model. It is therefore the very nature of the psychoanalytic system that demands a change!

(3) A multifunctional model is also called for by the very structure of personality. Robert Waelder stresses the principle of multifunctionality on the following basis:

> Freud described the phenomenon (anxiety) both from the angle of the id and from that of the ego. This two sided consideration gives rise to the presumption that the same method might be adopted and applied to all psychic phenomena,

< THE DEDUCTIVE BIRTH OF THANATOS THEORY >

and that a double or multiple conception of each psychic action would be necessary in the light of psychoanalysis.[50]

In conclusion, the need for a multifunctional model of the dream derives from (a) the pronounced difficulty in understanding the traumatic dream without it; (b) its being demanded by the current general construction of psychoanalytic theory; and (c) the psychoanalytical model of personality structure.

How, then, are we to interpret the drive for monism in dream theory when Freud's theories tend to be multifaceted in structure? And how are we to explain the tendency of Freud and his followers to avoid the problem of nightmares, and post-traumatic dreams in particular? I should like to raise a cautious hypothesis, stressing that it does not pretend to provide a monistic explanation (as this would contradict the very spirit of my essay.)

As mentioned, dream theory never underwent a real modification by Freud or his followers, all of whom seem to have viewed it as the most perfect of all psychoanalytic cornerstones.[51] The reluctance to attempt a revision came to the fore in Freud's own reaction: "Let us leave this dark subject . . ."[52] Freud's difficulty in relating to traumatic dreams has no doubt influenced the entire field of theoretical dream research. It may further be speculated that the stagnation in dream research, relative to research in other areas of psychoanalytic theory, had its origin in the avoidance of free and independent discussion on dream theory ever since contradictions on the nature of the traumatic dream became evident.

In trying to explain the urge to retain a monistic element within a pluralistic theory, one might point to a conflict between the inclination to objective research independent of subjective human wishes and tendencies and the psychological drive for uniformity.

In criticizing philosophers who paint themselves the illusion of a monistic, universal and overall superstructure, Freud cites Heinrich Heine: "Mit seinen Nachtmützen und Schlafröcken stopft er die Lücken des Weltenbaus."[53]

Freud and his followers were not prepared to give up their monistic approach to dream theory, despite the fact that it was contradictory to their way of thinking on other matters. As a pragmatic scientist whose point of view was inductive, Freud recognized that there was an inevitable lack of harmony in every theory, including his theory, particularly in his caustic characterization of the philosophers with whom he identified so closely. Nevertheless, he exhibits the disguised wish for the actualization of a unifying harmony. This need for harmony—the search for a unifying, holistic and monistic method—expresses a

< Scientific Myths and Lacunae: From Thanatos to Logos >

psychological need in scientists as well as in philosophers, one that seems to go hand in hand with the keen desire for objective research which makes no concession to latent wishes. Perhaps it is this split that conceals the deeper reason for Freud's encapsulation of his dream theory within a monistic system, while his general theory is pluralistic.

One must conclude that the post-traumatic dream and neurosis were the only reasons for Freud's revising drive theory. Thanatos was supposed to solve this problem via the rear door, but failed to do so.

Max Stern has observed that it was particularly important for psycho-analysis, so largely concerned with the problem of neuroses, to examine anxiety—one of the most striking expressions of that affect. The same author points out that *Pavor nocturnus*, which is one of the most striking manifestations of anxiety in human life, has been treated with some neglect in the literature . . . *Pavor nocturnus* tends to be overlooked.[54]

In the foregoing, we have seen: (a) There is no clear theory which can explain the nature of the traumatic anxiety. (b) The function of both the post-traumatic neurosis and the post-traumatic dream remains unclear. (c) Post-traumatic neurosis has not been integrated into the general theory of the neuroses. (d) The post-traumatic dream has not been inte-grated into general dream theory. (e) We still have no clear and accepted conception of the meaning of 'trauma'. (f) The development of the dream theory was blocked in the first place due to the difficulty of Freud and his followers in dealing with exceptions to their theory. (g) Thus, the inability to solve the riddle of anxiety prevented a possible under-standing of trauma, and the difficulty in understanding trauma, in turn, prevented conceptualization of the post-traumatic neurosis and dream. Finally (h), the avoidance of questions raised by the traumatic dream prevented formulation of integrated dream theory.

Another reason for the problematic development of psychoanalytic dream research was Freud's implicit message that his dream theory was complete and should remain untouched.

The untouched problem of anxiety and the difficulty of dealing with exceptions like the post-traumatic dream has resulted in a long chain of developmental problems in the field of psychoanalysis.

If Freud intended to bring about a more comprehensive under-standing of psychoanalytic phenomena by introducing Thanatos into psychoanalytic theory, he did not reach his goal. For, in labelling Thanatos mute, he reflected its inability to broaden the scope of psycho-analysis and instead brought us to a dead end. Nonetheless, Thanatos as presented by Freud did serve another, albeit tacit, goal: it prevented invalidation of his drive theory.

Thanatos was never proved clinically by Freud; however, neither was

< THE DEDUCTIVE BIRTH OF *THANATOS* THEORY >

it ever disproved. Whatever does not fit into the pleasure principle may be considered as belonging to it. And, since Thanatos is mute, according to Freud, extended drive theory could not be disproved.

It is true that most of Freud's disciples rejected the Thanatos theory. Still, they did not fail to receive the message behind the theory. Indeed, they carried it out even more carefully than Freud—to the point where everything must be explained in accordance with drive theory. Therefore, although Thanatos was rejected by all except Melanie Klein and Jacques Lacan, Freud's followers changed nothing in dream theory because Thanatos protected it.

Classical dream theory remained formally intact until the appearance of self-psychology, even though it no longer fitted either ego psychology or object-relation theory. Paradoxically, the rejected Thanatos theory helped preserve the monistic dream theory.

Despite what has been said above in regard to the speculative Thanatos, Freud's claim that Thanatos is mute may receive fresh meaning if it is connected to a new conception of trauma. If we look at the post-traumatic dream as a prototype of Thanatos, in which psychic death is reflected by the arrest of intra-psychic activity, we may speak of a temporary psychic death—Thanatos—when the psyche works in closed circles, without any primary process productions. The real traumatic state, then, would be a catabolic process in the spirit of Freud. This description defines the boundary between the so-called traumatic state and actual traumatization. The blurring of the notion of trauma, which I have illustrated herein, thus represents a scientific symptom. It results from the difficulty in explaining phenomenon that are problematic for a theory that must be preserved by any means.

< SCIENTIFIC MYTHS AND LACUNAE: FROM *THANATOS* TO *LOGOS* >

7

How Logos *Arose Mythologically from* Mythos

It may perhaps seem to you as though our theories are a kind of mythology and, in the present case, not even an agreeable one. But does not every science come in the end to a kind of mythology like this? Cannot the same be said today of your own physics? — S. Freud, "Letter to Albert Einstein", 1933

Now that we have seen how Thanatos was brought into psychoanalysis, we must ask how the new combination of Eros/Thanatos influenced the development of the discipline. The previous chapter pointed to the use of scientific mythology in Freud's solution to the scientific problem of the theory of drives, but the new theory of drives, which included the death drive, gave rise to another new problem. Thanatos stood in Freud's way. An unconfirmed theory had been created that threatened to undermine Freud's intellectual worldview. Thanatos threatened to give rise to despair over human progress. Freud therefore saw the need to rehabilitate the standing of Logos, upon which progress depends. Freud's conscious intention in *Moses and Monotheism* was to provide a psychoanalytic explanation based on scientific findings for the development of Jewish monotheism. In what follows I show that Freud's unconscious intention was to provide a mythological foundation for the standing of Logos and to grant it extraterritorial status, one which places it in conflict with psychoanalytic theory. I shall try to demonstrate the distinct and independent status of Logos in psychoanalysis, as acquired through myth.

The inductive psychoanalytic method placed drives at the centre of human existence, and Eros ruled. Acquired human qualities were considered derivatives of the drives. Thus, thinking and reasoning were treated as products acquired on a detour from the path toward direct gratification. In the context of the inductive psychoanalytic method, Logos has no different status from that of any other characteristic. Nevertheless, it was treated differently. Why? In the first place, because psychoanalytic epistemology, as derived from its meta-psychology, is

< HOW *LOGOS* AROSE MYTHOLOGICALLY FROM *MYTHOS* >

by nature positivistic, independent of psychoanalytic theory. Secondly, because in Freud's later work on the psychoanalysis of culture Logos became disengaged from drive theory and thereby attained—as I shall endeavor to show—a mythological status. The turning point came when Freud introduced Thanatos into drive theory to solve problems for which drive theory had not found any answers. The death instinct was first mentioned by Freud in *Beyond the Pleasure Principle*.[1]

By introducing Thanatos, Freud did to psychoanalysis what, according to Plato, Zeus had done to the primordial human being.[2] The archaic human creature was a pure product of Eros, and the gods were helpless against Eros' drives. In the biblical myth, God feared that Adam and Eve might eat from the tree of life and live forever. According to Greek mythology, the erotic force in man overwhelmed the gods. Zeus decided to put an end to this force by cutting the human being in two. Freud did the same to his drive theory. From that time on, Thanatos has been antagonistic to Eros. After being cut in two, mythological man no longer fought the gods, concentrating instead on searching for his second half. Since the introduction of Thanatos, Freudian man has been fighting against himself. This is primary masochism according to the new drive theory.

Prior to the appearance of Thanatos on the stage of psychoanalysis, Eros was constantly forced to fight against opposing forces. Dualism was always inherent in the structure of psychoanalysis. The conflict took place mainly between the superego and reality. These were relative forces. In contradistinction, Thanatos is an absolute—and an even stronger force than Eros, one that threatens to neutralize it.[3] We are not dealing here with civilization and its discontents, but with the discontents of human nature. After all, civilization might be able to change if it were not for the interference of the mighty Thanatos.

After mythological man was cut in two, the two parts searched for each other with the aim of uniting in an eternal embrace. This longing of Eros for the eternal, static nirvana can in reality be identified with the death instinct. Eros leads to Thanatos. It was only after Zeus took pity on human beings and moved their genitals from back to front that they could satisfy their sexual desires and reproduce. Now man could turn outward and become a dynamic rather than a static creature, and mankind was saved from nirvana. But happiness had to be restricted. Since united Eros threatened to become nirvana, the restriction of happiness was the only way to make life possible. This is what Immanuel Kant meant when he said that the human being strives for consensus, but nature knows best what is good for him and therefore splits him.[4]

As the designer of psychoanalysis, Freud was aware of this point. This is why he created a duality that attends to all the vicissitudes of

< SCIENTIFIC MYTHS AND LACUNAE: FROM *THANATOS* TO *LOGOS* >

drive theory: first, between the libido and the ego-drives, then between the drives and reality and between the drives and the superego; and finally between Eros and Thanatos. Yet the wish for unification continues. It is already evident in dream theory: the existence of a monistic wish theory within the realm of a multifaceted, multi-determined theory. Something of the same kind happened to the Logos.

The Platonic myth demonstrates that duality both protects human beings from the disastrous consequences of unification and deprives them of their greatest wish. Unification is both dangerous as well as desired. In the debate between Freud and Romain Roland[5] concerning 'oceanic feelings', Freud's view represents fear of fusion, while fusion is the object of desire for Roland. What constitutes a phobic element for one is the source of religious feeling for the other.

In psychoanalysis, Eros is in continual struggle and conflict with other forces; however, it never gives up trying to realize its final destiny—unification. From the moment that Thanatos was introduced, the hope that this destiny could be attained was destroyed forever. Where Eros is Thanatos neutralizes it. Similarly, the prescription for Thanatos is Eros itself: arrival at the static position. In this respect, Freud remarked: "It seems that the pleasure principle is in the service of the death instincts."[6] One illustration of this from everyday life is post-coital depression, which patients sometimes describe as 'a little death'. It seems that the drives function in a closed circle: just as Eros leads to Thanatos, Thanatos leads to Eros. Thus, while cutting mythological man in two was an act of Thanatos in that it prevented unification, it also enabled reproduction and thereby the continuation of life, ergo Eros. Because the drives go in a closed circle, there is no hope for unification—or any progress (towards any goal.)

Freud's attitude toward history was one of enlightened scepticism, very much in the spirit of Voltaire's: "Plus ça change, et plus c'est la même chose."

Continuing this line of thought, Freud said:

> It may be difficult, for many of us, to abandon the belief that there is an instinct towards perfection at work in human beings, which has brought them to their present high level of intellectual achievement and ethical sublimation and which may be expected to watch over their development into supermen. I have no faith, however, in the existence of any such internal instinct and I cannot see how this benevolent illusion is to be preserved.[7]

With this statement Freud's pessimism reached its peak. From then on he searched for an alternative. His pessimism was the source of optimism; it followed Kant's idea that nature uses man's anti-social drives to help him develop his capacities. Leo Rauch parallels this idea with

< How *Logos* Arose Mythologically from *Mythos* >

Judaeo-Christian eschatology,[8] in which the mythos of progress becomes a permanent vision. Freud, on the other hand, joined those who believed in the myth of progress made possible by the notion of reason. While in this he does not differ from Karl Marx, John Stuart Mill or Kant, the optimism that Freud derived from his pessimism is of a special nature: not only does the Logos not develop organically from Freudian theory, it is there despite this theory—one based on unchanging drives. The mythical Logos appears as a clear and steady factor that leads us slowly, although surely, to our goal.[9] We can sense a turning point in *Beyond the Pleasure Principle*, for although Freud did not discard the death instinct, he did find a way out of the closed circle by relocating unity and progress from the drives to the Logos.

Having used the Platonic myth to demonstrate the need for unification and progress, Freud now used another myth to show how unification and progress, which couldn't be realized through Eros, could achieve their goal through Logos. The myth of choice appears in *Moses and Monotheism*.[10] Unlike the drives, Logos is monistic. It could succeed where Eros failed. The dominion of Logos meant the abolition of opposites and of the unconscious. Because it is not ruled by drives, and therefore not prone to extinction, nirvana and stasis, Logos can lead to unity and progress. Logos is a defence against the unreliable Eros and 'oceanic feelings', against fusion and the blurring of boundaries. It preserves boundaries and saves the identity of the individual.

Since it exists above the drives and remains independent of them, there is no danger that Logos might produce contradictions within drive theory.

Moses and Monotheism became the focus of endless interpretations and comments. His commentators might be classified into those who took interest in his interpretation of ancient Jewish history and those who chose to find out about his personal Jewish roots. It seems to have been a special attraction to connect Freud the atheist to his Jewish background. Freud gave them a case by writing on Jewish myth for the first and only time, after using (throughout the development of his psychoanalytic thought) European, but especially Greek, mythology as demonstrations for the unconscious. One has to ask oneself what motivated Freud to write on Jewish history at the end of his life. He was not sentimental and did not give up his atheism in bad times. He never stopped being the enlightened cosmopolitan till his last days.

He must have had some other motives for writing this piece of work. The historical critics of Freud's Moses run into the pitfall of taking his historical reconstruction too seriously and in considering in it Freud's main goal. Their reservations concerning his historical analysis should have been addressed to the historians on whom Freud based his ideas,

< SCIENTIFIC MYTHS AND LACUNAE: FROM *THANATOS* TO *LOGOS* >

Ernst Sellin and Edward Mayer. As in his anthropological writings, Freud, when dealing with history, chose selectively those works best suited for the purpose of the conclusions he was driving at. If one looks at *Moses and Monotheism* in the context of Freud's writings, especially his late ones, one may detect in it an allegory on the development of Logos as a universal idea.

It is a well-known fact that Freud was vehemently attacked for his version of Moses, which critics said expressed Freud's contempt for the Jewish religion. I admit that I have difficulty in understanding what was so denigrating about this work. If one reads it thoroughly, one finds that, in so far as Freud could sympathize with any religion at all, his sympathy was granted to the Jewish religion—even if in his own interpretation. As we shall see, he considered Judaism only one step below pure enlightenment. In fact, he believed that enlightenment has its roots in Judaism. What more could one wish from an atheist like him.

An analysis of *Moses and Monotheism* will illustrate why the Freudian Logos is mythical in nature, for Freud identified the same duality and polarization that exist in psychoanalysis in Judaism.

Freud's interpretation on Moses and the foundation of Jewish religion is a myth rather than an historical analysis. It is an allegory. Freud built his thesis on the hypotheses of the historians Sellin and Mayer, and he constructed the following story: During the reign of the 18th dynasty in Egypt and under the influence of the Sun God in Heliopolis, the monotheistic concept came into being. Aten was the only and single God. The Pharaoh Akhenaton turned this monotheism into a state religion. It was the first case of monotheism in history. After Akhenaton (around 1350 BCE), with the ending of the 18th dynasty, this monotheism came to an end. And here, says Freud, our hypothesis begins. In Akhenaton's kingdom lived a man named Tuthmosis, who believed in the religion of Aten. As the governor in a province on the border, he came in touch with a Semitic tribe. After losing hope in the development of monotheism in the kingdom, he compensated himself by converting this Semitic tribe to the new religion. According to Ernst Sellin's research, the members of the tribe turned against Tuthmosis, killed him and brought an end to the new religion. Edward Mayer claims that the tribe united later on with Midianite tribes in Kadesh (Sinai) and took upon itself their religion—the worship of the Volcanic God Yahweh. The leader of these tribes was named Moses too. Later on, the two united tribes invaded Canaan. The result of the fusion of the two religions was the incorporation of Tuthmosis monotheism into the Midianite religion. There was a long period after the defection from the religion of Moses during which no sign was to be detected of the monotheistic idea, of the contempt for the ceremonial or of the great emphasis on ethics. Still,

< How *Logos* Arose Mythologically from *Mythos* >

some traces of monotheism were kept latently. After a latent period the Midianite God lost his original character and more and more came to resemble the God of Tuthmosis. The characteristics of Aten's religion as described by Freud were monotheism, the fanaticism of the only God who would not accept any other Gods, the acceptance of death (in contrast with the Egyptian religion, which denies death and therefore denies reality), and the choice of the people of Israel by God. Freud claims that Jewish history is familiar to us for its duality: There were two groups of people who came together to form a nation; two kingdoms into which this nation later fell apart (Judea and Israel); two Gods; two religions; and two founders by the name of Moses. In describing the polarization between the Midianite Moses and the Egyptian Tuthmosis, Freud identified two historical peoples, two religions and, later on, two kingdoms (Judea and Israel). According to him, the Jewish religion developed out of polarization and conflict between the wild tribes of Sinai and those who lived in accordance with Akhenaton's moral code. The key to the human capacity for abstraction and intellectual development lies in the package deal that the people of Israel made with the monotheistic God. The Jews gave up sensory perception in return for reason, sensuality in favour of an abstract, unseen God. This was a fanatical deity who would not allow any other gods. "The idea of a single God means in itself an advance in intellectuality."[11] According to Freud, the people of Israel agreed to this contract because they received in exchange a God who preferred them over other people. In addition to increasing their self-esteem, they were promised protection, although not intimacy, by the unseen God. It would seem, therefore, that monotheism is like pure Logos in that it, too, is universal, fanatical and lacking in intimacy. It, too, renounces sensuality but provides protection. (Freud speaks of the need for the dictatorship of reason.) [12]

Following Auguste Comte, Freud claims that humanity reaches its highest level of development when humans are able to recognize reality, and this occurs when they no longer need animism or metaphysics. Freud identified Logos with this reality principle; he denied that there is life after death. According to him, Judaism reached a higher level of development than the ancient Egyptian religion because it does not concern itself with life after death. To him, the ancient Egyptians' preoccupation with the afterlife proved that they could not recognize reality. This myth relating the choice made by the people of Israel demonstrates the development toward the rule of the Logos and confirms its universality.

Why did Freud choose the myth of the choice in order to prove his thesis? Had he been interested in a balance between Mythos and Logos, between sensuality and the abstract, he would not have chosen this

< SCIENTIFIC MYTHS AND LACUNAE: FROM *THANATOS* TO *LOGOS* >

myth. He needed a myth with an element of sacrifice because he wanted Logos to become the only ruler, and for that to happen Eros had to be renounced. Nothing in drive theory supports Freud's mythological analysis about the way that Logos became independent. Drive theory talks about detours, not about total sacrifices or the abolition of sensuality. One must therefore conclude that the Freudian Logos is not derived from Freudian theory. He did not need this myth in order to show that humanity was capable of achieving advanced powers of abstraction. Hellenistic culture provides clear-cut evidence that Logos and Mythos, reason and sensuality, can coexist in an advanced culture. In fact, Freud himself remarked that "harmony in the cultivation of intellectual and physical activity, such as was achieved by the Greek people, was denied to the Jews".[13] Freud did not, however, conclude that the Greeks had found the right modus; he was *a priori* in favour of the dictatorship of reason.

The next question relates to the analogy between Jewish monotheism and secular Logos, to how the neurosis of monotheism was overcome in the transition to secular, conflict-free Logos. Freud saw in every religion, including Judaism, an expression of neurosis. Freud's explanation of why the Jewish religion is neurotic goes as follows: The people of Israel sacrificed sensuality when they accepted monotheism, but it turned out that the sacrifice was not so absolute. Israel became annoyed with the unseen, abstract God, and moved away from Him in favour of idols. After returning to Him, they repressed their aggressive feelings toward Him. Their guilt feelings led to the compulsive rules of Judaism. However, since pure Logos is conflict-free, the transition from neurotic religion to non-neurotic Logos is not explained. There is a missing link. Freud made a mythological jump here.

What distinguishes the Jewish religion from secular Logos is neurosis. This missing link between pure reason and Jewish monotheism constitutes the mental leap that Freud made in order to place Logos on a high and independent pedestal. Here we find the wish element in his theory. He needed to set aside one reliable value from the deterministic, Thanatic world that he built. This is what I mean when I claim that Logos arose out of Mythos.

Another question raised by *Moses and Monotheism* is precisely why we should axiomatically assume that monotheism is superior to polytheism. Even if we suppose that the capacity to unite different elements serves as an indication of a developed culture, the question remains whether Judaism, as represented by Freud, unites the different elements. Let us put the question of unification figuratively. Suppose that the Shield, or Star of David, is a condensation of two separate triangles representing the dualism of the Zoroastrian religion ("Shapes light

< How *Logos* Arose Mythologically from *Mythos* >

and creates darkness, makes peace and creates evil", Isaiah 45.7). Judaism's message was the abolition of dualism. But how can Freud claim that his interpretation of Judaism is a condensation of contradictory elements when the two religions of the two Moses, as he described them, are not put together through either synthesis or complementarity? In Judaism, according to Freud, unification is achieved through incorporation. Akhenatonic Judaism swallowed the Midianite religion: "The central fact of the development of the Jewish religion was that in the course of time the God Jahweh lost his own characteristics and grew more and more to resemble the old God of Moses, the Aten."[14]

It is difficult to speak of 'unification' when one element is consumed or destroyed. Neither monotheism nor the monistic Logos created by Freud expressed the broad spectrum of human possibilities that derive from clinical psychoanalysis. If this is so, then his monism leads to restriction, and we can hardly speak of progress.

This problem of unification and synthesis can also be illustrated in terms of parenthood. According to Freud, religious feelings are connected to the emotional wish for a parent. Jewish monotheism implies preference for the father over the mother. The incorporation of the Midianite religion into that of Aten means giving up one parent— the mother. There is no place in Freud's theory for the joint rule of both parents. Either the mother rules or the father does. It is a fight in which one side must always have the upper hand. As its culture developed, Judaism gave up the mother. Thus, pure Logos won the battle against Eros, against conflict, neurosis and Thanatos.

Among the paradoxes inherent in Freud's thesis as presented in *Moses and Monotheism* is that Logos, too, has its own Thanatos. Freud says that the day will come when science will have found a solution to every problem.[15] Freud believes that we are not only on our way to the truth (Popper), but that we will reach it, and that this will mean the end of history and of science. This is the Thanatos of the Logos. However, there is no end to history in the world of the unconscious. In saying that, Freud buries not only history and science, but psychoanalysis itself. The uniqueness of psychoanalysis lies in the picture it paints of the unconscious world, and if the bottom line of psychoanalysis were that pure Logos should rule, what would it all be about? "Where there was id, there shall be ego" is the death sentence of psychoanalysis. So Logos leads to the Thanatos of the fascinating 'underworld', as well as of the discipline that exposed it.

Although Freud seems to have achieved his desired goal, the hegemony of the Logos, he was not completely satisfied. He claimed repeatedly that Logos must be imposed compulsively. If it cannot dominate naturally, but only by means of compulsion, then perhaps it did not

< SCIENTIFIC MYTHS AND LACUNAE: FROM *THANATOS* TO *LOGOS* >

ultimately succeed in escaping the drives. Compulsion is a term used to indicate neurosis. If Logos must function as a policeman, there is something defensive about it. As we have seen, Freud never integrated Thanatos into his dream theory. Thus, the dream theory remained monistic—and logos is not the only monistic element in the psychoanalytic discipline. For the monistic Eros in the dream (wish-fulfilment) comes face to face with the monistic existence of Logos. Thus Logos must not be the only ruler after all.

Not only did Freud try to explain the special status of Logos in a mythological way, and not only is there a missing link in his explanation, but he himself doubted the totalitarian and autonomic existence of the Logos he created. This can be concluded from the defensive way in which he treats the Logos.

The Freudian autonomous, all-powerful Logos is a mythological creature. The Logos arising from Mythos closes a circle. This is because Logos was considered superior *a priori* in Freud's epistemological methodology. It was only in the late stages of his work that Logos acquired its special status within the context of psychoanalytic theory. The end supplied the evidence for the beginning. The circle closes.

The two myths described herein are antithetical to one another. They express the struggle between Freud's Hellenist *Weltanschauung* and his vision of the Jewish philosophy of history: a circular or sinusoid conception versus a linear conception of the flow of time. Fathum versus choice, stasis versus dynamics, nirvana versus eschatological vision, determinism versus indeterminism—draw yourself a picture versus do not draw yourself a picture; an ahistorical approach versus an historical approach. The Hellenistic concept of Freud is bound to a vision of the unconscious that possesses neither a past nor a future. The myth of unification and duality is taken from Plato's *Symposium* and represents Hellenism as if it were seen through Freud's prism. It appears in *Beyond the Pleasure Principle*, one of his early anthropological writings. The myth of choice in *Moses and Monotheism* represents the antithesis of that myth and seals Freud's writings. It represents the conscious world. These two myths reflect the intellectual and emotional struggle that shaped Freud's psychoanalytic thinking.

We can easily see that, although Freud considered myth inferior to Logos in his anthropological philosophy, he made massive use of the former as a cornerstone of his theory. The Eros/Thanatos motif follows like a scarlet thread through all his late writings, although from time to time he tried to break the circle and move in the direction of the version of Jewish linear history, toward progress and teleology and vice versa. Although the oscillations continued throughout the later writings, they

< How *Logos* Arose Mythologically from *Mythos* >

never succeeded in tipping the scale. In opposing the *Moses and Monotheism* allegory as the last opus representing the Logos, Freud deemed the last opus representing the Eros/Thanatos cycle "terminable and undeterminable".[16] In so doing he expressed the *Weltanschauung* of Empedocles, from whom he borrowed the idea of the two instincts and in whom we also find the idea of the moving wheel of the instincts.[17]

The Hellenistic part of Freud pulled him toward stasis, destiny and tragic man, but, in contradistinction to Nietzsche's solution, for him it is not the mythical Eros who stands by humans in their mythical eternal flight and enables then to pull themselves up, but the human Logos.

In *Beyond the Pleasure Principle* Freud speaks of predestination and about the demonic human being. And, as antithesis, the last opus is written as an allegory to the opposing vision, saying 'no' to the statement "The destiny of life is death."[18] Scientific future and the reparation of the world do exist. But this does not change the fact that the myth of Sisyphus, or Aristophanes, or the philosophy of Empedocles, are much closer to deductive clinical Freudian psychoanalysis than the myth of choice that represents the Logos—a position outside the struggle of the instincts. It is interesting that Spengler's *The Decline of the West*[19] was published three years before *Beyond the Pleasure Principle*, for both were published in a period when history seemed to have been trapped in a *cul de sac*. Spengler's sinusoidal philosophy of science reflects the decline of the West, a recapitulation of the fall of Rome. His idea of the eternal return is a variation on Empedocles' theme. The exposition of my thesis may represent an analogy to the book of Ecclesiastes. The author of Ecclesiastes produced a *deus ex machina* toward the end that Freud echoed in *Moses and Monotheism*, for he declares: "The end of the matter: all has been heard. Fear God, and keep his commandments; for this is the whole duty of man. For God will bring every deed to judgment, with every secret thing, whether good or evil."[20] And Freud says, in so many words: "The end of matter: all has been heard. Respect science; for that is all that man has."

In contrast to Ecclesiastes, however, Freud continued to be overwhelmed by doubts. He oscillated between the two contradictory visions until the end of his life. After all, 'mythology' is a combination of myth and Logos, and it originally meant spiritual and artistic creativities, which were closed to poetry as well as to science and philosophy.

< Scientific Myths and Lacunae: From *Thanatos* to *Logos* >

PART IV

The Epistemological Split in Psychoanalysis

Part IV discusses how far the psychoanalytic discipline, which deals with the irrational, went in its tendency toward the veneration of Logos. We examine the reasons for this phenomenon and its consequences for the field. Chapter 8 deals with the dictatorship of the Logos and the epistemological double split in psychoanalysis. Chapter 9 treats the unconscious elements that compelled Freud to deal with psychoanalysis from a positivistic perspective, despite the fact that it is anti-positivistic in its essence.

The dictatorship of the Logos brought about the formation of epistemological lacunae. We examine two such lacunae: the refusal to attend to the idea of complementarity, which exists in inherent fashion in psychoanalysis; and the difficulty and delay in the development of the idea of countertransference. Chapters 10 and 11 treat these two topics respectively.

8

The Dictatorship of the Logos

Motto: Our best hope for the future is that intellect—the scientific spirit, reason—may in process of time establish a dictatorship in the mental life of man.
—S. Freud

Where did psychoanalysis take Logos? This chapter treats the dictatorship of psychoanalytic Logos against the background of the development of science in the twentieth century. Freud wished for dictatorship of reason, while physicists like Pauli considered irrationality a part of reality, including science.

"The dictatorship of reason" was a phrase coined by Freud.[1] While it is obvious that eighteenth-century Enlightenment leaders and nineteenth-century positivists were advocates of reason, they did not need to talk in terms of dictatorship. Freud's motive for using this metaphor is of particular interest since his theory accords the material such a high place. It is also of interest because he referred to the dictatorship of reason in "The Question of a *Weltanschauung*", where he denigrated *Weltanschauung* because of its dictatorial features. In that article Freud cites Heine's verse in which the poet mocked philosophers for trying to construct a hermetic world structure,[2] and yet he made a similar declaration when he said that "the best hope for the future is that intellect—the scientific spirit, reason—may in the process of time establish a dictatorship in the mental life of man".[3] It seems that his desire for an all-out war by the ratio was influenced mainly by new winds blowing in the twentieth century, winds which Freud considered dangerous.

Yet Freud began by attacking philosophy, certainly not a new wind, for the deductive system that it represents presents a perfect and hermetic harmonious worldview. As such it might be likened to a hat that is forced on the head. While philosophy is a substitute for religion, science represents reality and realism, and corresponds to external reality. Science is aware of the need to separate knowledge from anything suggestive of illusion or that which is desired out of emotional

< THE DICTATORSHIP OF THE *LOGOS* >

need. Psychoanalysis, on the other hand, cannot form a *Weltanschauung* worthy of the name because it is incomplete. And yet Freud did admit that he had a *Weltanschauung* when he said: "A *Weltanschauung* created upon science has, apart from its emphasis on the real external world, mainly negative traits, such as submission to the truth and rejection of illusion."[4] Here he manoeuvred himself into paradox, for he was fighting *Weltanschauung* by means of a *Weltanschauung*.

In Freud's eyes science differs from philosophy because it realizes the incompleteness of the knowledge at its disposal. Nevertheless, he was a utopian in assuming that science could reach completeness— "Everything that interests us finds its fixed place"[5]—particularly since he used similar wording to show that a *Weltanschauung* is out of place in our culture: "A *Weltanschauung* is an intellectual construct which solves all the problems of our existence uniformly on the basis of one overriding hypothesis which leaves no question unanswered."[6]

Freud's writings seem to demonstrate considerable solidarity with the same philosophy that is the subject of his attack. In speaking about the dictatorship of reason, he himself may be suspected of applying emotional motives; for example, when he expresses his concern that the world, which he believes in, is under threat. In short, he seemed to desire the same perfect reason, without lacunae, as the philosophers he was attacking. Here we should remember what Kierkegaard had to say in this context: "The denial of the subjective leads to self-contradiction, for even the most abstract proposition remains the creation of human beings."[7] Freud rejected *Weltanschauung* as a kind of monism, yet his own rationalism constitutes a form of monism. Even after coming to terms with the philosophers, he continued to attack intellect as it manifests itself in modern physics—which he called another *Weltanschauung*. For, in his view, modern physics forms a counterpart to political anarchism because relativity "went into their head".[8] Freud accused the physicists of claiming that there is no such thing as clear knowledge of the external world or of scientific truth, neglecting to mention Niels Bohr and Werner Heisenberg, even though his "The Question of the *Weltanschauung*" may be indirectly related to the Como congress of 1927, where Bohr announced complementarity as a general, pan-scientific epistemological idea.[9] For Freud this was a kind of scientific suicide, for if scientific truth were to lose its absolute meaning, it wouldn't matter which view we adopted. Every view would be as good as any other. Physical and mathematical paradoxes do not lead us anywhere. Freud might lead one to conclude that he was referring to a group of madmen. But when his paper was published in 1933, this scientific 'anarchism' was already an integral part of the *Zeitgeist*.

Paradoxically, the relative decline of rationality during the twentieth

< THE EPISTEMOLOGICAL SPLIT IN PSYCHOANALYSIS >

century was based on rational reasons: beating the Logos with Logos.[10] The tendency to accept subjectivity as part of science was not a result of romantic motives, but the rational acceptance of our lack of rationality. In this context, Fritjof Capra says that every time a twentieth-century physicist answers a question, he ends up with a paradox. Paradoxes are context-bound. By definition, they contradict the idea of the *tertium non datur*—the exclusion of the third possibility.[11] And, in the spirit of mathematical paradoxes, P. W. Bridgman says: "The problem of how to deal with the insight that we never get away from ourselves is perhaps the most important problem before us. This is an insight that the human race deliberately refuses to admit."[12]

The discoveries of modern physics caused huge changes in our concept of space, time, matter, object, cause and result. They overthrew Newtonian concepts of separateness, for space and time were no longer separate dimensions in the New World. It was in this changing world of the twentieth century that psychoanalysis paradoxically developed. For the discipline that uncovered the centrality of the irrational in human beings is the same discipline that uses all possible means to defend pure rationalism. As early as 1916, Freud declared: "Where there is id, ego shall be."[13]

Like Francis Bacon, Freud viewed subjectivity in science as a problem that could and should be eradicated from scientific observation. Thus, psychoanalysis became the great champion of pure rationality and positivism—of Logos. At the same time we are witness to the interesting phenomenon of distinguished physicists being preoccupied with irrationality in science. Each side seeks its counterpart.

Quantum mechanics evoked, in some of the leading physicists and other natural scientists of the twentieth century, a deep interest in philosophy, psychology and psychoanalysis, mainly from the epistemological point of view. Among those who expressed ideas about irrationality in science and the unconscious were Niels Bohr, Werner Heisenberg, Pascal Jordan, Wolfgang Pauli and Max Delbrück.

Wolfgang Pauli wrote a series of works on epistemology and psychology, including the scientific and epistemological aspects of the unconscious. According to Pauli, the natural sciences are unthinkable without irrationality. Irrationality is an integral part of reality because comprehensive thought includes pictures and symbols that are connected to the rational aspect of reality by complementarity. In fact, what he is talking about here is primary-process thinking in science (*Aspekte der Ideen vom Unbewusten*).[14]

Werner Heisenberg believed that what we observe is not nature itself but nature exposed to our method of questioning. Thus, he argued that atomic physicists cannot adopt the role of detached objective observers

< THE DICTATORSHIP OF THE *LOGOS* >

because they become involved in the world they observe in that they influence the properties of the objects observed.[15]

According to Bohr, quantum theory reveals the essential interconnectedness of the universe. Although matter is composed of particles, we cannot deconstruct the world into its independently existing smallest units because these particles are not basic building blocks in the sense of Democritus and Newton, but merely idealizations. In his view, particles are abstractions that can only be observed and defined in reference to their interaction with other systems.[16]

In order to understand more about the dictatorship of reason in the context of the psychoanalytic discipline, we must first deal with the double split in psychoanalysis. Our point of departure is that while psychoanalysis is revolutionary in content, it is conservative in structure. It is revolutionary in that it brought the unconscious world into the centre of the anthropological picture, and conservative in that it tried to mould this revolutionary content into a positivistic structure.

The split between the meta-psychological and the clinical levels in psychoanalysis was first identified in the 1960s. George Klein[17] and Roy Schafer[18] discussed this split from the analytic point of view, while Paul Ricoeur[19] and Jürgen Habermas[20] discussed it from the philosophical point of view. It appears that, from the outset, there is in the discipline of psychoanalysis a gap between meta-psychological assumptions and clinical psychoanalysis. Analytic meta-psychology, as shaped by Freud and his disciples, pleads for a science built according to strict positivistic criteria, a science that functions as an objective tool with which one can observe the unconscious. Clinical psychoanalysis—unknowingly—developed along a different path, for the simple reason that the basic concepts of clinical psychoanalysis do not lend themselves to a positivistic philosophy.

But, there is a second split in psychoanalysis—on the interclinical level. The clinical theory suggests that analysts observe the rule of abstinence during sessions. The long-held view that the analyst is a blank screen or mirror begs for a positivistic translation from the meta-psychological to the clinical-technical level. On the other hand, psychoanalysis also observes the rules of overdetermination[21] and multiple function,[22] neither of which can be considered positivistic; indeed, they indicate the contrary.

The split between the meta-psychological and the clinical levels is less serious than that on the interclinical level. After all, it is possible to develop a clinical theory that overlooks meta-psychological postulates, considering them rudimentary.[23] But the interclinical split, with its positivistic influence on the discipline, presents a very serious problem.

< THE EPISTEMOLOGICAL SPLIT IN PSYCHOANALYSIS >

Long after the meta-psychological postulates were openly or tacitly rejected on the part of the psychoanalytic community, the split clinical theory remained. True, there are those who have expressed discontent with the clinical theory, but the question is whether and how this discontent has resulted in changes in this theory. In the following chapters I shall deal with the deep reasons for this epistemological split (Chapter 9), why it was not realized for so many years, and the consequences of the 'scientific symptoms' of denial (Chapters 10 and 11).

< THE DICTATORSHIP OF THE LOGOS >

The Psychoanalysis of Epistemology

9

Engagement with the dictatorship of reason and with the epistemological double split in psychoanalysis brings us to question the motives that may underlie epistemological approaches. The main questions are: How could such a split occur and how could the psychoanalytic community bear it for so long a time before bringing it out into the open?

The Hebrew Encyclopaedia, inspired by idealistic philosophy, defines epistemology as:

> The scientific discipline that desires to understand consciousness and to make the special tie between knower and known the subject of a higher consciousness . . . consciousness is always connected with the knowing subject. Nevertheless, it is imperative to distinguish between consciousness as a psychic process and consciousness as the fruit of this consciousness. Denying this distinction leads inevitably to faulty psychologism. Epistemology is interested in the value of truth and its relationship to the structure of knowing consciousness, and is not interested in the psychic context. The moment we deny the value of truth in respect to consciousness, science becomes impossible.

According to this definition, psychology and psychoanalysis should not have anything to say about epistemology or science, as the latter two belong to a truth that is independent of human motives.

The *Encyclopaedia Britannica* and the *Encyclopaedia for Social Sciences* both define epistemology in relation to knowledge. If, however, we translate the word 'knowledge' into Hebrew (*da'at*), we can see that it does not bear the cognitive connotation attached to it in the Anglo-Saxon definition. For the biblical *lada'at* (to know) carries the connotation of a deep relationship between people, with an erotic element. Thus, in Hebrew, the sharp dichotomy between object and subject does not seem relevant, and consequently the sharp division between a philosophical study of consciousness (= epistemology) and the study of consciousness in its psychological sense becomes less obvious.

< THE EPISTEMOLOGICAL SPLIT IN PSYCHOANALYSIS >

In the twentieth century we have witnessed a growing demand for a science of science—a meta-science that would replace philosophy as the queen of sciences—one that includes philosophy, sociology and psychology. Returning to the psychoanalysis of epistemology (in general), our analysis of the epistemological position of psychoanalysis is just one example of the contradictions that disciplines, especially those in the humanities and the social sciences, may become embroiled in when they try to preserve the classical epistemological concept.

The double epistemological structure in psychoanalysis is just a private case. For anyone who assumes a dichotomy between object and subject is of necessity behaving in accordance with double standards. Thus, every historian has the illusion that he or she is offering an objective description. This indicates that another epistemology underlies the positivistic structure, and what such historians do not express explicitly may be inferred from reading between the lines. This characteristic is found in every science and produces a split between the true and the false self of science.

Psychoanalytical epistemology is an allegory *par excellence* of what we can see in other sciences. In the psychoanalytic process we can clearly see what generally occurs in interactions between object and subject and is generally overlooked.

One of the basic questions in which epistemology is involved concerns the certainty with which humans can be secure in regard to their recognition of the world. (*Episteme* means security). The need to know clearly *what really is* brings with it the tendency to look for laws, harmonies, determinism. Alongside the need for this kind of security, however, there is a curiosity about the new, the urge to wander in the field of the unknown. This need for security and curiosity are basic human tendencies, and both are necessary for human existence. One cannot do without them. Psychoanalysis assumes that the more we wander in our internal world through fantasies and associations, the more we are able to understand ourselves. Such understanding is a precondition for our understanding the external world. In this regard, allowing ourselves to 'get lost' leads indirectly to more security. Despite this tacit assumption, however, classical psychoanalysis chose an epistemology that limits the possibility of getting lost.

If we accept that there are revolutions in science, then psychoanalysis was one of them.

In science as well as in history, revolutions tend to provoke counter-revolutions, or at least attempts at counter-revolution. These revolutions result from curiosity and a daring to seek out the unknown. The counter-revolutions represent the need for security and conservatism. In the case of psychoanalysis, the counter-revolution comes

< THE PSYCHOANALYSIS OF EPISTEMOLOGY >

from the inside rather than the outside. In a way it is a self-goal: the revolutionary aspect of psychoanalysis consists in its foregrounding of the very laws of the unconscious, and the counter-revolution is represented by classical epistemology, which forced itself on this internal world and stood in the ways of its expressing itself.

Freud's attack on philosophy and on modern physics presents us with a starting point for a *Weltanschauung*. He excluded the possibility of a *Weltanschauung* in the enlightened world because in his eyes philosophy is characterized by speculation and wishful thinking, whereas science is true to reality and refrains from idealizing. In other words, science does not allow itself to fall prey to the subjective and philosophy, which can be accused of behaving in many ways like a religion.

Freud was probably not aware that his own idealization of the dictatorship of reason raises the suspicion that he posited the need for such a dictatorship out of fear that his own *Weltanschauung* would be crushed. Nor did he realize that he was demonstrating solidarity with the philosophy he attacked when he called for a science in which everything finds its place, a science based on reason and independent of the subjective.[1]

Expressing these ideas in 1933, Freud attempted to bring us back to an organized Newtonian world where space and time are well defined and there is a clear separation between physical masses. By expressing the wish for a harmonious, intact and deterministic world which is predicted and has a clear-cut direction, he preserved classical epistemology in the face of evidence from the clinical theory to the contrary. Freud's struggle to retain the Newtonian worldview in his new discipline ignored all the changes that had occurred since the time of that theorist.

So Freud had a *Weltanschauung* at the same time that he denied having one. He represented himself as a revolutionary scientist who dared to go wherever scientific observation led him, and who created something out of nothing. But his need for a structured scientific worldview revealed a security measure that dictates the *Weltanschauung* he denied.

Although he stressed the importance of individual autonomy—we may almost say autarchy[2]—by claiming that he could bear reality without needing God or any other artificial support, when it came to paradoxes, reality became too confusing. Thus, his concluding a section on anarchism by saying, "I have neither the desire, nor the capacity for going into this more deeply . . . ", [3] seems to indicate that he refused to deal with matters that threaten the secure world in which he believed.

Bertrand Russell represents the antithesis of Freud. His world is a

< THE EPISTEMOLOGICAL SPLIT IN PSYCHOANALYSIS >

hurly-burly place that contains pleasant as well as unpleasant things, in a random order. In his view, desire shapes it into a reasonable system that might counteract the fear of open spaces, from a kind of agoraphobia.[4] Carl Schorske claims that modernity does not define itself as either originating in or opposing the past, but as independent from the past.[5] He seems to be describing the perfect revolution, and in the same spirit as Bertrand Russell. Some historians of culture, however, claim that every revolution reflects two contradictory vectors. The one directed toward the future represents the energy that overpowers the gravitational force of conservative elements. The second vector comprises these conservative forces.

In line with the proposition that something cannot be created out of nothing, Gerald Holton has the following to say about the theory of relativity:

> The so-called revolution which Einstein is commonly said to have introduced in 1905, turned out to be at bottom an effort to return to a classical purity. The relativity theory shifted the laws of space–time from the sensorium of the Newtonian God to the reasoning of Einstein's abstract *Gedankenexperiment*.

Holton contends that a revolution in science is not generally an iconoclastic act of destroying the existing scientific basis, but a turning toward the ideal past, which expressed itself in few and uncomplicated hypotheses.[6]

Holton seems to be describing Freud's kind of revolutionary, a person with deep roots in the past and a phobia for certain scientific and cultural developments. For Freud's scientific daring went hand-in-hand with a counter-phobic process. Thus, while he was the first scientific revolutionary of the twentieth century, he had a need to restrain his creation so that he would not lose control of it. Freud also warned of the nihilistic consequences of revolutions in other fields, for example, in physics. Freud belongs to the group of revolutionaries who do not feel completely at ease with their innovations, and who try to keep them within bounds by employing positivistic constructions. Thus, he oscillated between, on the one hand, belief in the victory of reason, and, on the other, the fear of being swept away and swallowed up by the very world of the unconscious that he had revealed. Thus, too, he declared that wherever there is id, there shall ego be, but he tempered this statement by adding "the ego is not master in its own house".[7] Lewis Feur concludes that Freud was a revolutionary of the unconscious who turned his energy from a potential political revolution toward a scientific revolution. In this connection, he notes that the motto Freud gave to his *Interpretation of Dreams*, *acheronto movebo*, means "If I cannot bend the powers above, then I will move those of the underworl."[8]

< THE PSYCHOANALYSIS OF EPISTEMOLOGY >

The unconscious world charmed Freud, but he had to distance himself from it in order to moderate its implications. In *The Future of an Illusion*, he discusses new scientific ideas in terms of development and advancement, not in terms of revolution, hesitating to call himself a revolutionary.[9] That was too dangerous. But this does not explain his need for the dictatorship of reason. In criticizing the oceanic feeling of Romain Rolland as "a feeling without measure and limits", Freud showed that he was not able to detect in himself a feeling that would bind him to the outside world. There is nothing safer than the self-feeling.[10] Martin Wangh tried to find a dynamic explanation for Freud's stand towards irrationality and fusion. His starting point is Freud's attack on Romain Rolland's suggestion to consider 'oceanic feelings' as fundamental to religious experience. Wangh comments that "it is puzzling that Freud, who, more than any other man, dared to look into the 'wider world' of psychic life, including his own, felt so much stress from Rolland's challenge". In *Civilisation and Its Discontents* Freud objected to the idea of oceanic feelings in an awkward way. He rejected the idea emotionally:

> I cannot discover this oceanic feeling in myself. It is not easy to deal scientifically with feelings . . . If I have understood my friend rightly, he means the same thing by it as the consolation offered by an original and somewhat eccentric drama-tist to his hero who is facing a self-inflicted death—We cannot fall out of this world. That is to say, it is a feeling of an indissoluble bond, of being one with the external world as a whole.[11]

According to the documentation worked out by Wangh, Freud's father was a weak, ineffectual man, while Amalia Freud was an over-whelmingly powerful, sexual and possessive person; though Freud adored and depended on his mother and yearned to approach her for the satisfaction of his need, he could not help fearing, avoiding and defying her. He was formed by ambivalent feelings toward her. She loved him as her narcissistic extension. His Oedipal fears toward his father were in large part a defence against matriarchal dread. He was fleeing from fusion with the maternal object. Death was linked for him with the woman, the mother, the goddess of death. So far Wangh. From all this it is obvious why Freud claimed in Civilization and Its Discontents that "there is nothing of which we are more certain than the feeling of our self, or our own ego. This ego appears to us as something autonomous and unitary, marked off distinctly from everything else."

Martin Wangh claims that oceanic feelings were too frightening for Freud. This is why he was unable to give in to his urge for an unlimited bond with his mother and tried to overcome it by turning to the ratio-nality of his father.[12] Extending Wangh's thesis, it seems to relate to

< THE EPISTEMOLOGICAL SPLIT IN PSYCHOANALYSIS >

Freud's warning against counter-transferential falling in love, against falling in love in the reality (which is considered by him a pathological phenomenon), as well as against anarchism in science, all of which imply the limitlessness that results in loss of the ego and suicide. In the analytic situation, analysts are threatened by unlimited, boundless feelings toward their patients. This necessitated a very clear separation between patient and analyst to prevent that dangerous side effect of counter-transference from taking over. This fear of blurred boundaries led to denial of the reciprocal situation in analysis and to repression of the unconscious world of the analyst, whom he assigned the role of mirror.

Clinical experience, however, has shown that the less analysts fear blurred borders, the more they are able to fulfil their task. The fact that Freud needed very defined borders for himself in his dialogue with the world does not necessarily mean that this is the case with other analysts. Although we analysts are aware that Freud based the fundamentals of psychoanalysis solely on his own personal, cultural and clinical experience, his influence on other psychoanalysts was such that generations of analysts have suffered from splits between their private and public theories, or they have inflicted splits on the movement whenever discrepancies surface between them and the clinical world. The loyalty to Freud has been too strong to allow analysts to go beyond their own possible fears of the unknown.

That the split in psychoanalysis is not solely Freud's problem is evinced by the fact that seventy years have passed without the issue having been raised in the psychoanalytic literature. Another indication of underlying causes external to Freud may be found in the fact that ego psychology flourished and led psychoanalysis in a positivistic direction for so many years. The hegemony of ego psychology was indicative of the need to broaden classical epistemology to include an emphasis on the independent status of the Logos. For ego psychology and all that may be wrong with it is an interpretation of Freud's psychoanalysis that accords with his epistemological concept.

Freud's conflict between his self and his science, between falling in love with the unconscious that threatened to overpower him and his wish to keep these oceanic tendencies (which he denied) under strong control, was echoed in ego psychology. But, by abnormally enlarging the reality aspect in psychoanalysis, ego psychologists were endeavouring to prove that the conflict between security and free-floating search had been settled.

There are three epistemological approaches to looking at science: the mystical approach, the intersubjective approach, and the one most

< The Psychoanalysis of Epistemology >

familiar to those in the West—the objective positivistic approach, which dichotomizes between object and subject. The objective approach may also be broken down into models. The aim of the model that has been common in the West for the past three hundred years is for the subject to occupy the object by penetrating it and thereby making it its own.

According to this approach there is a special relationship between the knowing subject and the known object, which represents two opposing poles with a special tension between them. It is based on the urge of the subject to penetrate the object, conquer it and make it part of oneself. The objective approach, which began with the Enlightenment, pits man against nature, man against reality. Man is supposed to conquer nature and reality for himself. The attitude is one of tension and confrontation, or what Abraham Maslow has called "defensive science".[13] Clearly, psychoanalysis does not fit into this objective epistemology.

The second model of the objective approach was that of ancient Greek science, which was also based on a dichotomy between object and subject, but, contrary to the modern Western approach, it observed nature from a distance, with respect and perplexity. The dominant psychological mechanism at work in ancient Greek science was idealization. It is like love from a distance. Thus, the ancient Greeks conceived nature as an indivisible and viewed this unit as a work of art that could not be touched or dissected without being harmed.[14]

According to Maslow the mystical approach to science is grounded on fusion between the subject and the object, between the observer and the world. In his view, Taoist science is not defensive and neurotic like Western science, because the relationship between the scientist and nature is a natural one and not one of tension resulting from the fear of being overcome. The Taoist does not stand and view the world as an observer. He or she feels at one with nature. This feeling of fusion tends to frighten modern Westerners.[15]

The third epistemological approach to science—the intersubjective approach—represents a compromise between total fusion and remaining completely separate. Its adherents believe in the subjective nature of human existence, and therefore they do not believe either in a strict division between subject and object or in complete fusion. Among these who have adopted this approach are the dialogical philosophers Herman Cohn, Franz Rosenzweig and Martin Buber. According to Hugo Bergman, "The synthesis does not engulf the thesis and the antithesis." Instead of synthesis, the intersubjective approach posits that, while the two entities preserve their separate existence, there is a relationship between them that creates a sort of cocoon around them.[16]

Although psychoanalysis has never characterized itself as dialogical, it is epistemologically similar to the dialogical or intersubjective

< THE EPISTEMOLOGICAL SPLIT IN PSYCHOANALYSIS >

approach, because analyst and patient both preserve their individuality and at the same time form a unit together. This is a very complicated position from the psychological point of view. To be in touch with the world without being sucked into it, and to be in touch with oneself without losing oneself as Narcissus did, is not an easy proposition. As Devereux warns, "The deeper one reaches into objects, the more the phenomenon one intends to study becomes attenuated until eventually it vanishes altogether."[17]

But those who strive to fuse with the world run the risk of sinking into the ocean. The intersubjective approach tries to prevent both the attenuation of which Devereaux warns and the split of positivism.

< THE PSYCHOANALYSIS OF EPISTEMOLOGY >

Complementarity in Psychoanalysis

This chapter treats one of the results of the dictatorship of psychoanalytic Logos. The concept of complementarity, which constitutes one of the cornerstones of psychoanalysis, may be considered a lacuna. For many years, the idea is not mentioned anywhere in psychoanalytic literature, and when it is mentioned, it is obfuscated and distorted.

Complementarity as was first defined by Niels Bohr, and, leaning on the uncertainty principle of Heisenberg, is a monistic idea which assumes the possibility of experiencing an object or an idea in different ways. According to the original physical definition, the different alternatives are expressions of the same object, although they do not appear at the same time and may contradict each other. In quantum mechanics there are endless possibilities of the same order.[1] According to the less rigid definition by Klaus Michael Meyer Abich, complementary phenomena don't necessarily contradict each other, and the *sine qua non* of his definition is that they do not appear at the same time.[2]

If one were to limit oneself to what psychoanalysis has had to say about complementarity, this would be a very short section. Complementarity was not mentioned in the literature until 1967, when Devereux claimed that Freud was the father of the idea of complementarity and not Niels Bohr. The existence of a symptom assumes that its meaning is unconscious. According to Devereux, the moment that meaning penetrates the consciousness, symptoms disappear. In his view, Freud intended to say that there is a complementary connection between the unconscious and the symptom. He deduces this from the original German which speaks of "Verhältnis von Vertretung". Strachey translated this as "inseparable relation", while Devereux claims it should be "the relationship of reciprocal representation". This would mean that, as long as the symptom exists, it is not possible to throw about the unconscious material that underlies it. Since the symptom disappears the moment that the unconscious becomes conscious, the relationship between them is complementary.[3]

< THE EPISTEMOLOGICAL SPLIT IN PSYCHOANALYSIS >

To my way of thinking, it seems questionable to credit Freud with the idea of complementarity on the basis of the translation of a single term. Freud was generally very precise in his expression. If he intended to make use of complementarity in psychoanalysis, he would have mentioned it explicitly. In fact, complementarity does not fit his philosophy.

It is my intention to show how the few analysts who have related to the notion of complementarity since Devereux have utilized it, and what we might deduce about the resistance to the very idea.

In 1977, Pinchas Noy discussed complementarity as a framework for a multi-model in psychoanalysis. Noy's basic hypothesis is that the uniqueness of psychoanalysis is in its meta-psychology, which is based on a multi-model theoretical system—a system that is composed of several theoretical models. The fact that psychoanalysis is not based on one theoretical model is not a weakness, but a source of power.

When a represented object is complex, it might not be possible to abstract it into one model and different models are usually needed in order to cover all its aspects. Freud, however, did not mean to construct a multi-model system. He hoped to find a monistic model. He was repeatedly obliged to revise his theoretical models and to add new ones in order to cover the new clinical phenomena which he discovered. As psychoanalysis is a system within which every clinical phenomenon has to be described according to all possible points of view, it has to allow for the addition of new models without disrupting the whole system. Thus, psychoanalysis developed into an open and flexible system.

The psychoanalytic multi-model as exposed by Noy is built on the principle of complementarity. He puts together the models (economic, structural, genetic, topographic and dynamic) under the Bohric umbrella of complementarity. But here the problems start: The basic assumptions of the meta-psychology are positivistic in their nature (they are based on Newtonian assumptions), while complementarity is anti-positivistic by nature.[4]

Noy's attempt to describe a positivistic multi-model within a complementary framework represents a mix of two contradictory epistemologies. However, where one mixes epistemologies, *tertium non datur* should be the rule. One cannot employ an epistemology in which there are no clear borders between objects to explain a multi-model based on such borders. A multi-model cannot describe psychoanalytic theory coherently if it does not reflect the intersubjective situation. In other words, Noy's multi-model deprives complementarity of the fact that the principle of uncertainty—its main element—is an essential quality of the world.

Arnold Modell takes a different approach to complementarity. He

< COMPLEMENTARITY IN PSYCHOANALYSIS >

suggests the coexistence of two epistemologies, reflecting two different psychologies—one- and two-person psychology. The processes of one-person psychology answer to the rules of science, and those of two-person psychology answer to the rules of hermeneutics.[5]

Modell's two epistemologies and two psychologies are inseparable; they cannot be placed in two different categories. According to him, it is not logical to consider analysts as mirrors or blank screens since they communicate with their patients. On the other hand, analysts are blank screens in the sense that they act as transferential figures, facilitating the bringing up of contents connected with introjected images (one-person psychology). Modell therefore proposes

> that we accept the opposing and contradictory views of the epistemology of psychoanalysis. Each view carries with it its own assumptions, models and metaphors . . . We require a new advance of thought—the epistemological device of complementarity, which permits us to accept without striving for synthesis the two opposing traditions of natural science and the science of the mind.[6]

Here again we witness the idea of complementarity, this time turned into a technical device aimed at combining two incompatible elements. Thus, Modell does not accept complementarity as deriving naturally from the structure of the discipline. For him it is nothing but a trick.

Heinz Kohut proposes that complementarity exists between the psychology of conflict and self-psychology. He praises Freud's scientific objectivity and, at the same time, offers "a more broadly based objectivity than that based on the 19th century scientist—an objectivity that includes the introspective emphatic observation of the participatory self". Kohut seems to have been charmed by Bohr's idea, seeing the transition from the epistemology of classic physics to that of quantum physics as analogous to the transition from macro-relationships (the Oedipal complex) to the molecular investigation of micro self-psychology. To him, the existence of an emphatic introspective observer defines the psychological field.[7] Thus, he seems to recognize the meaning of complementarity for psychoanalysis. He recognizes the existential crisis of the twentieth century and aims to alleviate it through self-psychology. When it comes to epistemological formulations, however, Kohut still oscillates between the old and the new. He cannot relinquish the idea of 'scientific truth' and has difficulties in defining the entity 'broad spectrum objectivity', which he himself coined:

> It is by no means necessary for analysts to discard the nineteenth century ideal. Under specific circumstances it is as valid now as it was then . . . There are certain depth-psychological observations, which, via their observation, influence the field they observe . . . in the vast majority of instances, the depth-psychology obser-

< THE EPISTEMOLOGICAL SPLIT IN PSYCHOANALYSIS >

vations will exert no more influence on the field he observes than the physicist who observes the movement of the stars.[8]

Kohut is fascinated by the new physical theory, which reminds him of the transition from classical psychoanalysis to self-psychology, but he still hesitates with respect to the idea of objectivity. He does not feel altogether comfortable with classical objective truth, but does not see how he can develop a new epistemology without losing stability and security. He oscillates at the edge of positivism.[9] Thus, Kohutian empathy does not deviate from the classical epistemological entities.

Noy, Modell and Kohut have a lot in common. To some extent, they are all enamoured of the new epistemology of the twentieth century, but all of them remained poised on the brink—an ambivalence that is especially clear in Kohut's writings. And none of them can accept the far-reaching consequences of complementarity, in which analyst and analysand are seen as functioning in one field.

This is not a matter of division between micro- and macro-relationships (Kohut), between one- or two-person psychology (Modell), or between metaphysical content and the anti-positivistic structure (Noy). Despite the fact that the psychoanalytic community tended to disregard complementarity or to misinterpret it, complementarity is reflected in its very essence. That the precursor of the idea of complementarity is implied in psychoanalytic theory can be demonstrated easily through two examples: ambivalence and the principle of multiple determination.

Ambivalence is a basic psychic structure that derives from contradictory drives. The capacity to live with some ambivalence is necessary for the intra-psychic homeostasis and for wrestling with the world. This is why analysts endeavour to avoid viewing problems in either/or terms. The negation of the *tertium non datur* becomes affirmative in the unconscious world in which negation does not exist but is derived from the function of the ego.

A second precursor of complementarity in psychoanalysis is the principle of multiple determination, which Roy Schafer connects to the epistemological problem. Schafer, related different and sometimes contradictory meanings to the same phenomenon as a principle central to psychoanalytic interpretations. Thus, he views the time devoted to confronting the analysand with the need to see things in black and white as overdetermination.[10]

Neither ambivalence nor multi-determination accord with the positivism of psychoanalysis. Overdetermination is contradictory to the general theory of the neuroses, to the meaning of symptoms and dreams. This, however, does not suffice to prove the immanent states

< COMPLEMENTARITY IN PSYCHOANALYSIS >

that are reflected by psychoanalysis via complementarity. An excellent example in this regard is the interrelationship between primary and secondary processes. Both are integral parts of the same psyche. Both processes function constantly even though we generally witness the activity of only one of them at a time. Thus, although the reality factor remains outside the picture in the dream state, and only the primary process is at the scene, we have the ability to put aside the primary process when awake, while the psyche is oriented toward the outside world. Certain phenomena let us know that the primary processes are always there, however, such as slips of the tongue.

Primary and secondary processes seem to be incompatible, and yet they work hand in hand in the service of the psyche. They cannot be disconnected one from the other. The silence of psychoanalysis in respect to the principle of complementarity stems from the fear of shaking the foundations of the positivism, a disturbance that Freud feared might turn into scientific anarchism. As an antithesis to positivism, Paul Feyerabend offers a scientific anarchism. Feyerabend's scientific anarchism means the following: everything goes. He claims that our knowledge is essentially ideational. As this is the case the history of science is chaotic and full of mistakes. There is no rule or law in science that was not disproved at one time or the other. The big inventions occurred only because some thinkers either decided not to be bound by certain methodological rules or because they unwittingly broke them. Attempts to attack the basic ideas of science evoked taboo reactions that do not differ basically from taboo reactions in so-called primitive societies.

If science is irrational in its assumptions, its methodology, its proofs and its rejection of new ideas, why not try a new way—anarchism in science. Feyerabend suggests that the results will be better than what was achieved in the old way.[11]

An intersubjective approach accepts neither positivism nor scientific anarchism. Just as it does not accept human knowledge independent of the interaction between subject and object, it is not in favour of 'everything goes'. The question of what should be considered compatible or incompatible is decided on the basis of interaction. This means that no psychoanalytic interpretation would be valid unless it resonated in the unconscious of both analyst and patient. When Freud considered an epistemology, he conceived of the one derived from quantum physics as an expression of scientific anarchism, which he feared would lead to every explanation being considered right or wrong at the same time.[12] However, in claiming that a dream can be interpreted in several different ways according to the principle of overdetermination in psychoanalysis, and that every interpretation is valid within the given

< THE EPISTEMOLOGICAL SPLIT IN PSYCHOANALYSIS >

associative context, I do not mean that every possible interpretation will be correct in every context. Within the framework of the context, there are binding rules. Thus, while many psychoanalysts admit that dreams should be interpreted in context, this does not imply that they accept anarchy.

The reasons why complementarity was resisted in psychoanalysis are hinted at between the lines. Complementarity may be viewed as a monism for the weak. We see the mountain and the paths leading to the top from only one side. We may never see the other side. By recognizing complementarity we would be giving up the possibility of understanding. We must be aware of our limitations.

In line with the idea that existence is a unity that cannot be grasped, Efraim Katzir advises us to be humble.[13] Henry Edelheit brings the complementarity of the weak to clear expression when he states that "Psychoanalytic psychology and neurophysiology are interconnected only by a complementary principle which does not suppose any inclusion, interaction, reciprocity or injection of the terms and concepts of one language into the other".[14]

His complementarity, like that of Katzir, is nothing more than an artificial technical solution. Because it does not connect anything, complementarity loses its meaning in his definition. Complementarity connected to the duality of psyche and body is not in a position to produce any new ideas. All it says is that psyche and soma are inseparable and that they cannot be translated one into the other. This does not accord with Freud, who believed that it would some day be possible to translate psychic language into biological terms. This aspiration for the unity of body and soul goes hand in hand with understanding (and not with mysticism).

Complementarity might also imply that we are not on the road to the truth that Freud said could be reached and that Karl Popper claimed we could only know the way to.[15]

Loss of determinism means loss of control and sinking into a fallacious world. As long as one refuses to accept the principle of indeterminism, one may claim that, although understanding has not been reached, the effort to attain it is worthwhile because there is a chance that one will reach it. Complementarity has a weakening effect. Some even claim that it reeks of theology. Indeed, it may be time for theologians to employ complementarity when they seek to bridge the inability to grasp divinity and its ways (modern theologists often claim that there are many ways to God). The worst that could happen is that we would find that God is not rational, but someone who plays with dice. Paradoxical as it may sound, if complementarity is associated with both theology and nihilism, the achievements of the Enlightenment are

< COMPLEMENTARITY IN PSYCHOANALYSIS >

cancelled. Moreover, intersubjective complementarity suggests the possibility that the analyst/analysand relationship may be an inflictive one.

It is much easier to continue deciphering the psychoanalytic narrative via the classical deterministic way. The parameters are more accurate and 'scientific'. Should we risk losing all this? So psychoanalysis keeps looking back to the world of yesterday, seeking the good old order. Dealing with the wild unconscious, with its threat of obliterating borders and incipient nihilism, seems to invite special precautions and the search for a secure epistemology. And complementarity, with its paradoxical simultaneous representation of nihilism and unrest of the inner world, on the one hand, and monism and order, on the other, is the sort of unitarity that is not ruled by the ratio of the nineteenth century. If we accept complementarity as leaning in the direction of vitalism,[16] which does not accept a materialistic interpretation of the world, are we not perhaps implying a return to the romantic period? To the existence of a mysterious power?

Resistance to a new psychoanalytic epistemology that includes the notion of complementarity may also stem from the fear that this will lead to a loss of the analyst's authority. The recognition of a semi-permeable membrane and of intersubjectivity removes the authoritative element from the psychoanalytic situation. The analyst will no longer be able to reside in an ivory tower, protected from the analysand. He or she will be more vulnerable, will lose the benefits of psychoanalytic infallibility. Countertransference will lose its special status. Indeed, Alice Miller claims that, in transposing trauma as a reality to a component of drive theory, Freud was endeavouring to defend parenthood from the accusation of child seduction. According to her, the shift to drive theory placed the exclusive blame for such fantasies on the child's 'dirty mind'. So does the Oedipus complex result from the child's imagination. In taking this approach, the analyst identifies with the parents and thereby prevents the child (analysand) from confronting them.[17] Applying Miller's idea to psychoanalytic epistemology, the basic approach of the analyst would be to distance him or herself from the analysand.

Psychoanalysis illustrates how complementarity benefits science. The discovery that thinking takes place simultaneously on the conscious, subconscious and unconscious levels provides a multidimensionality that gives the process of thinking its creativity: it allows for tree planting between the conscious and the unconscious, between the inner world and the outer world, between the subjective and the objective, between the sphere of picture and symbols, that of words and thoughts. In other words, it is a prescription for creativity.

< THE EPISTEMOLOGICAL SPLIT IN PSYCHOANALYSIS >

Max Delbrück argues that the Cartesian cut between observer and observed, between inner and external reality, between mind and body, is based on the illusion that the physical world has no subjective component. In experiencing the mental world, we include a wider repertoire of perceptions than when experiencing the physical world. Both worlds are complementary one to the other. And, to Bohr, a good science is one that works according to these same principles. That there is an integral connection between complementarity and creativity is also evinced by the fact that adults who preserve a sense of childhood by thinking in modes of complementarity and not being bound by rigid rationality are the ones who produce the most original ideas.[18]

By denying complementarity its proper role, psychoanalysis denied an idea inherent in the nature of the discipline. When Freud exchanged religious monotheism for secular monism, he was not aware of monism's intrinsic contradictions of clinical-analytic principles.

Bohr extended the idea of complementarity from physics to biology, endeavouring to explain its necessity through the "killing off mechanism". If one wants to know what goes on inside cells *in vivo*, he claims, one is forced to destroy the cells, but then they are no longer *in vivo*— they are dead.[19] He defended the idea of the "killing off mechanism" on the concrete level of biology. While his thesis is open to dispute, this idea should be tested on a higher philosophical level. The *Tu quoque* argument claims that whenever one tries to push an idea to the end, one ends with a tautology and thereby kills off the object of the thought. The moment we try to turn the object into a something moribund by excluding intersubjective and the context, we destroy the experiment of thought. What we end up with is dead and not the live thing which we set out to explore.[20]

If we transpose this idea to psychoanalysis, we can say that we shall never know what is really going on inside the other if we do not view what comes to light in an interactive context. If we fail to do this, all we can expect is moribund.

Complementarity also goes hand in hand with the psychoanalytic principle of uncertainty. We are never in a position to fix the psychoanalytic process in a deterministic manner. We never know at the beginning whether our exploration will lead us to what really happened in the past. In shifting from reconstruction to construction, from attempting to reconstruct the past to attaining subjective truth that accords with the transference relationship, we give up the idea of the objective truth.

We have seen that the lacuna around a discussion of complementarity in psychoanalysis was complete until the 1960s despite the fact that it has always been a part of psychoanalysis. It did not have any

< COMPLEMENTARITY IN PSYCHOANALYSIS >

place in theoretical epistemological dialogue. Since the 1960s, complementarity has been disputed, but mainly in order to show that it does not abolish the positivistic approach. Thus it is clear that the epistemological split is still in operation.

< THE EPISTEMOLOGICAL SPLIT IN PSYCHOANALYSIS >

11

Countertransference as a Lacuna in Psychoanalysis

This chapter deals with the development of the idea of countertransference in psychoanalysis—specifically, how its existence was denied over time, and what theoretical-clinical price was paid for this denial. Although the phenomenon of countertransference was known to Freud himself, many years went by before it passed openly into the psychoanalytic theoretical framework. Countertransference is the most conspicuous example of a lacuna in psychoanalysis. This lacuna is a direct result of the dictatorship of Logos, a dictatorship that was accepted without objection by the psychoanalytic community. This passive acceptance led in turn to a significant retardation of the development of the psychoanalytic clinic.

The vicissitudes of countertransference in psychoanalysis during the hundred years of its existence seems to recapitulate those of the epistemology of science. The reaction in psychoanalysis was delayed because philosophers and physicists introduced subjectivity into science before psychoanalysis became the guardian of Logos. There is no better way to follow the epistemological split in psychoanalysis than to trace the development of the concepts of transference and countertransference. For what may seem at first glance to belong to the theory of technique, in fact reflects the psychoanalytical conflict over epistemology.

The main question here is whether the relationship between analyst and patient is characterized by a thin partition, a semi-permeable membrane, or any membrane at all. Is the analyst a mirror? What flows from one side to the other? Is analysis a one- or two-sided process, a one- or two-person psychology, or both? These questions are the core of the transference–countertransference question that clinical psychoanalysis has been trying to answer for so long. The transference–countertransference relationship extracts conscious and unconscious memories, feeling, conflicts and relationships. There is no better way to discover in depth how a relationship develops than by following this process.

Freud never claimed that the analytic process produces transfer-

< Countertransference as a Lacuna in Psychoanalysis >

ence. It is, rather, a universal phenomenon which is merely revealed and heightened in analysis.[1] As meta-theoretical scientific factors forced themselves on the development of the theory of technique and on the clinical attitude, they brought about a distortion of clinical observations that prevented the conclusions that might have been derived from these observations. Neglect of countertransference as a factor in the analytic situation inhibited the formation of optimal observations on many aspects of the analytic session. The epistemological changes that took place over time followed the clinical observations in a reversal of the process that took place in the beginning phase of psychoanalysis, when Freud's clinical attitude was shaped by the epistemological vision.

The first obstacle in the paradoxical development of transference—countertransference is connected with Freud's self-analysis. By engaging in self-analysis, Freud not only turned Bridgman's saying that we can never get away from ourselves[2] upside down but also acted counter to his claim that the subject of analysis could never be its object. In the act of self-analysis Freud put himself above himself by creating a psychic parthenogenesis. For him the analytic dyad was a basic requirement, and nobody after him saw in self-analysis a replacement for analysis. Peter Gay, when referring to Freud's auto-analysis, cited a letter from Freud to Paul Schilder saying that those people among the first generation of psychoanalysis who never underwent analysis were not proud of it. Referring to himself, Freud hinted that he might ask for himself the right for a special status.[3] Thus, the same man who built the foundations for a theory in which transference relations are a cornerstone started off with an act that contradicts those foundations.

The indirect but very clear message that he transmitted to his followers was that he was the one valid exception to a rule that nobody else should ever break.[4] It was as if Isaac Newton had claimed that the physical rule of gravitational forces did not apply to himself. Strangely enough, this situation and its impact on the development of psychoanalysis has never been discussed in the literature. Or perhaps not so strange, for it appears that Freud's position on this matter, and its blind acceptance by his followers, strengthened the taboo on open criticism, a taboo that paralyzed free thinking not only in this matter but in general. Freud's act of psychic parthenogenesis invested him with the position of a superman who is to be adored for his supernatural force. The foregoing is not meant to deny Freud's originality. Nor can we forget that he was much braver than his followers in daring to expose aspects of his self-analysis to demonstrate the force of psychoanalysis. But why has nobody ever asked the crucial question: How could Freud,

< THE EPISTEMOLOGICAL SPLIT IN PSYCHOANALYSIS >

who was never analysed, handle his countertransference while analysing his patients and colleagues, and what analytic consequences may his blind spots due to self-analysis have had on these analyses and on subsequent generations of analysts and analysands?

Although Freud believed in the neutral position of the analyst, it was he who introduced the concept of countertransference. In "The Future Prospects of Psychoanalytic Therapy", Freud claimed that counter-transference obstructs progress in psychoanalysis.[5] As he saw it, countertransference begins with the analyst's emotional response to stimuli emanating from the patient. This response results in an influence on the part of the analysand's unconscious on the unconscious feelings of the analyst. Freud's formulation places the 'responsibility' for this situation on the analysand. It is always what the patient evokes in the analyst, never what the analyst might evoke in the patient. This is similar to what he says happens in the Oedipal situation, where it is always the fantasy of the child, never the fantasy of parents or what this might evoke in the child. The parents are protected by Freud's drive theory. The analyst is protected by his definition of countertransference. Freud viewed countertransference as an obstacle to analysis, one that analysts should cure themselves of through self-analysis. In a later version, it was to be handled in the training analysis.

It is of interest that the only time Freud went into clinical detail about countertransference was connected with falling in love. In his "Observations on Transference Love", [6] he warns about the danger of countertransference love, calling it a pathology: "[If transference love] seems so lacking in normality—this is sufficiently explained by the fact that being in love in ordinary life is also more similar to abnormal than to normal mental phenomenon."[7] In his view, falling in love is always pathological, and he warns against it; transference love is even more so. It is always a sign of resistance of the patient: "For the analyst this situation is an unavoidable consequence of a medical situation."[8]

Here Freud was hiding behind his medical gown. His only mention of transference love in the context of the countertransference reaction that this might evoke in the analyst has a special meaning. Was Freud never annoyed, jealous, outraged or humiliated in his relationship with his patients? In discussing countertransference reactions, why did he choose the possibility that the analysand would fall in love with the analyst? It is not illogical to conclude that this represented the greatest danger to himself because it meant breaching the borders of his self and thereby his feelings of complete sovereignty. In choosing this example, Freud was warning himself and us against the oceanic feelings that counteract the feeling of the neutral position and mastery. The conse-quences are clinical as well as epistemological, and the greater the

< COUNTERTRANSFERENCE AS A LACUNA IN PSYCHOANALYSIS >

danger for the analyst, the stronger the warning. One who is so afraid of his own feelings that he must constantly warn himself against them runs the risk—despite self-analysis—of severe repression and consequent blind-spots that will hinder his ability to relate freely to analytic material.

We may conclude that three factors prevented Freud from seeing more deeply into the complex problem of countertransference: (1) his positivistic scientific approach, which imposed itself on his clinical observations; (2) his fear of oceanic feelings and his need to counteract them; and (3) his need to protect the parental generation, and thereby the analyst, from the patient. (The idea of protecting the parent comes from Alice Miller.)[9]

These factors did not reflect only Freud's idiosyncratic sensitivities. Many of his generation of analysts accepted the positivistic approach either because they too were raised on it, or because this philosophy also served them as a rationalization against their own fears. They too were in need of protecting themselves as parent figures in transference. But the fear of oceanic feelings cannot be considered universal. Many analysts, for example, fear their aggression toward the patient much more.

That Freud's warning struck a chord in the unconscious of several generations of analysts may be seen in the forty-year period that elapsed before the theory and clinical experience of countertransference—one of the most important issues of psychoanalysis—began to be developed in the psychoanalytic community. Horacio Etchegoyen wrote:

> The 40 years that elapsed between Freud's discovery and the time when this theme was studied again, cannot be said to have passed in vain, yet it is certain that nothing substantial was contributed to the study of countertransference during that time . . . during the first half of the century the theory of counter-transference does not appear in the theory of the technique. Science, says Kuhn, evolves through crisis . . . I really think that something like this happened with the recognition of countertransference at the middle of the century.[10]

Etchegoyen's statement leads to two questions: First, how could this repression continue for forty years of psychoanalysis? And, second, was the change or turning point that he says took place in the 1950s really as revolutionary as he claims? Before relating to these questions, let us see what actually happened.

Freud left his followers the message that countertransference is a negative phenomenon. James Kern expressed this when he compared the relationship between countertransference and psychoanalysis with that between an inflamed wound and the surgical act.[11] Analysts are expected to overcome the subjective as reflected in countertransference

< THE EPISTEMOLOGICAL SPLIT IN PSYCHOANALYSIS >

in a manner similar to the way in which Francis Bacon suggests scientists can overcome subjectivity by isolating it.

Freud's definition of countertransference underwent many modifications over time. As is the case with trauma, these definitions reflected a scientific symptom: the enormous conflict between deductive psychoanalytic theory and inductive clinical experience. They reflect even more the analysts' emotional difficulty in admitting the extent to which they may be involving themselves in the analytic process, and their fear that this will shatter their impregnable neutral position. For, even when it was recognized that the phenomenon of countertransference belies the possibility of maintaining a strictly objective position, the theory of psychoanalytic technique continued to view it as something that could be handled objectively. The very need for a special term to define this emotional position reflects the need to place analysts in the unique position of neutral observer. Indeed, the concept of therapeutic neutrality, which reflects the positivistic attitude, is based on the classical conception of the scientist who observes from the outside.

The Freudian attempt to guard analysts from involvement and contamination eventually led to the creation of the myth of analytic neutrality that is not a part of Freudian theory. In other words, while Freud wished to prevent contamination in the transference relationship, he didn't mean to restrict or limit the analyst's field of activity while doing so. This myth has caused several generations of analysts to become less spontaneous in their private lives as well as in the analytic dyad, and it has seriously influenced their social and political involvement. The positivistic attitude has led these analysts to deny what would seem to be obvious on the basis of simple common sense as well as from clinical psychoanalysis: that every deed or omission of the analyst influences the process. In failing to recognize this simple truth, analysts deny inter-activity and thereby inhibit the therapeutic process. It is difficult to accept that these defence mechanisms are due solely to the wish to remain faithful to positivism. Indeed, it is quite clear that the inhibitions on the part of analysts are related instead to their fear of exposing aggressive and libidinal wishes toward their analysands. The rationalization that neutrality is necessary for scientific reasons has provided an excellent excuse for many analysts who confidently refrained from checking themselves.

The first time that a Freudian analyst wrote about countertransference in a positive light was in 1949, when Paula Heimann discussed it as an analytic tool, rather than a danger.[12] Heimann claimed that analysts should utilize their emotional reactions during the analytic process in order to better understand the dynamics of their patients and to handle the process in a more differentiated way. She defined

< COUNTERTRANSFERENCE AS A LACUNA IN PSYCHOANALYSIS >

countertransference in a broad way by claiming that it covers all the feelings, that the analyst experiences toward his patient. She considered the analytic situation to be a relationship between two people, which was at the time quite a new concept. Heimann claimed that it is not the aim of the analyst's own analysis to turn him into a mechanical brain which can produce interpretations cut off from his emotions, but to enable him to sustain his feelings in order to subordinate them to the analytic task. The emotions roused in the analyst will be of value to his patient, provided that they are used as one more source of insight into the patient's unconscious conflicts and defences. Heimann's contribution and those of Heinrich Racker, [13] Margaret Little[14] and others led to a gradual change in approach, albeit not without considerable resistance at first.

It took another seventeen years until the next change in theory occurred with the appearance of George Devereux's *From Anxiety to Method*.[15] Devereux further broadened this theory in his 1978 *Ethnopsychoanalysis*.[16] In these two works, Devereux shows analysts and other social scientists how to turn subjectivity into a positive force, and how to use psychoanalysis to accomplish this. He does so by positing that the social scientists, who view subjectivity solely as a source of error, should take a leaf from the books of the psychoanalysts, who treat it as their main source of information. According to Devereux, analysts who are in tune with the beneficial aspects of countertransference allow themselves to be disturbed by its effects and then analyse these effects on themselves. In other words, analysts understand their patients to the extent that they understand the effects the patients create in them. Interaction means that both halves of the dyad are at the same time observers and objects of observation. Each side observes at the same time in two different ways—the analyst from the outside and the patient from the inside. Both are not possible at the same time, and they may result in different and even contradictory information. But they complement each other and result in a more complete observation. They are part of one unit. (This is only one of the ways in which Devereux introduces complementarity into psychoanalysis.)[17]

With regard to the borders between the two objects, Devereux cites an analogy that Heisenberg drew in connection with the principle of indeterminacy:

> One cannot reach into the electron, not even reach it at all, without creating a situation which radically differs from the one which one intends to investigate, since the indeterminacy is at the surface of the electron. The locus of the disturbance is therefore also the locus of the partition, so that the outer boundary of an object is not an a-priori datum, but the product of inspection.[18]

Thus Devereux does not consider the ego as having borders, but as

< THE EPISTEMOLOGICAL SPLIT IN PSYCHOANALYSIS >

being a border—one that changes in accordance with the nature of the inspection. "All behavioural science experiments are either rigid sticks or loose stick experiments, and the manner in which the stick is held is determined by one's theories, and in turn radically influences them."

Devereux's epistemological concept is summed up in the following statement: "The nature of the knower is the *sine que non* of the nature of the known."[19] His post-modernist visions of psychoanalysis eliminate the subject/object dichotomy. Since Devereux, countertransference has become more than a useful instrument in psychoanalysis. Having been recognized as the main source of analytic understanding, subjectivity has become an inherent value in psychoanalysis. However, Devereux's clear exposition of his epistemology, including its pan-cultural applications, is rarely cited in this context.

Roy Schafer and Donald Spence made important contributions to the application of the new epistemology. Schafer showed how ego psychology damaged psychoanalytic development and he suggested a new language of psychoanalysis.[20]

Spence developed the concept of narrative truth to replace that of historical truth, and suggested that construction rather than reconstruction be the goal in analysis. This entails giving up the pretension that we can ever achieve a reliable reconstruction of the past and concentrating instead on analysing the analytic relationship as a means for cure.[21]

Eveline Schwaber developed the concept of the "unity of the organism" to investigate the common field in which patient and analyst communicate. She claims that psychic experience is not the "property of the individual", but rather the property of the more inclusive system of which the individual is a part. In this situation, we cannot know things in themselves:

> We can only know in a patient's inner world that which we perceive. The world we perceive must include ourselves as perceivers. So, too, the patient's perceptual world must be understood in the context of his perception of us; we can only know that perception insofar as it includes us, rather than a phenomenon that goes on independent of us.[22]

With the sort of perception Schwaber outlines, the entire perspective of psychoanalysis changed. As Arnold Cooper puts it, this calls for a change from the diachronic to the synchronic view. In other words, instead of dealing with analytic material in terms of historical development, analysts should define analytic material in terms of what it means in the here and now, ignoring the question of how they got that way.[23]

The new epistemology has brought with it a change of emphasis with regard to the importance of infantile neurosis in the psychoanalytic

< COUNTERTRANSFERENCE AS A LACUNA IN PSYCHOANALYSIS >

process. In the new concept, infantile neurosis is viewed more as an underprivileged set of current fantasies rather than as historical fact. And transference resistance becomes the core of the analysis, to be worked through mainly because of the rigidity that it imposes on patients, rather than for any important secret that it may conceal. This new epistemology does not merely imply that the analytic process involves a two-person psychology, or that this psychology involves participation—even major participation—on the part of the analyst. What it makes clear is that this participation is basic to the analytic relationship and process. This means that there can be no dichotomy between transference and non-transference relationships, as this would imply that analysts determine from the outside, on their own, when patients distort and when they do not. The new epistemology rejects the argument that the analyst's view is necessarily the correct one. Truth in the new epistemology is multi-determinable.

Object-relational analysts are not good candidates for the new epistemology since they define countertransference as everything the analyst does or brings to the relationships[24] and consider countertransference positive only if it facilitates the analyst's commitment.[25] Their concentration on the role of the analyst, including the possibility of countertransference neuroses,[26] does not indicate their willingness to accept this epistemology.

Joseph Sandler states that the reactions of analyst and patient cannot be completely detached from one another. For him, externalization, projections, projective-identification and the patient's putting parts of his or her self in the analyst, can neither describe nor explain the dynamic processes of transference and countertransference. By assuming that a complicated system of unconscious cues must exist,[27] Sandler represents those object-relationalists who recognize that something is missing in the analytic approach, but are still far from a new epistemology.

Analysts who can be identified as exhibiting the new epistemology are those who, as James McLaughlin puts it, place the analyst's personality and behaviour, and the patient's awareness of these, squarely in the centre of the analytic process, in other words, those who define countertransference as everything that they do or bring to the relationship.[28] Spence, Schafer and Schwaber are representatives *par excellence* of this approach.

One might come to the conclusion that self-psychologists, with their emphasis on empathy as the main tool in analytic work, are anti-positivistic enough to accept the new epistemology. But, refined as the empathic experience may be, it still involves the effort to see oneself in the shoes of the other person from the position of observer rather than

< THE EPISTEMOLOGICAL SPLIT IN PSYCHOANALYSIS >

participant. In this respect the theory of empathy does not contradict classical notions of the object—subject relationship.[29]

Barry Protter defines empathy as existential or experiential knowing which is monadic by nature in contradiction to contextual or interpersonal knowing, which encompasses what may be called knowing oneself and the other within the context of a dyadically constructed interpersonal relatedness.[30] In my view, monadic empathy can be compared to a giraffe whose neck can reach the far side of a fence but whose legs remain on familiar ground. Thus, while Kohut may have thought that he arrived at a new epistemology in psychoanalysis, he was mistaken.

While the new epistemology began to be recognized during the 1980s, those who follow it are still a minority, and its opponents are rife. Charles Hanly represents those who cling to classical Freudian theory: "We are not always able to find the meaning, but it is there to be found, independently of any pattern-making activity on the part of the analyst. The task of interpretation as Freud conceived it is to make the interpretation correspond with the operative unconscious wishes of the dreamer—wishes that have a definitive nature of their own.[31] He interpreted the idea of theory-based observations as rationalizations for countertransference resistance to the threat posed by the drives—the instinctual unconscious.

Returning to the delay in recognizing the inter-clinical epistemological split in psychoanalysis, there does not seem to be any logical explanation for the lapse of forty years between the 'discovery' of countertransference and its recognition as a positive analytical tool. It would seem that the idea permeated the consciousness of analysts a long time before it began to be accepted. Etchegoyen describes how Racker caused such unease when he presented his ideas that he was told that an analyst to whom "these things" happened should analyse himself.[32] In my view, statements such as these prove that Etchegoyen was wrong in viewing the work of Racker and Heimann as revolutionary in the sense that it caused a shift in the psychoanalytical paradigm. The change in attitude toward countertransference did not bring about a change in epistemology because we have seen that one can believe in the positive value of countertransference and still consider himself an objectivist.

Robert Stolorow posited that psychoanalysis is unique among the sciences because it is the only science in which the observer is also the observed.[33] But this is the case wherever more than one person is involved in any sort of interaction. Psychoanalysis is, however, the science that can demonstrate this type of interaction in the most refined way. This is why it is so strange that it took psychoanalysis such a long

< COUNTERTRANSFERENCE AS A LACUNA IN PSYCHOANALYSIS >

time to discover inter-subjectivity. The denial of the countertransference phenomenon and its implications for the epistemology of psycho-analysis seems to me to be the biggest and most striking of the lacunae in psychoanalysis.

< THE EPISTEMOLOGICAL SPLIT IN PSYCHOANALYSIS >

PART V

Evolution, Revolution, Revelation and Creativity in Psychoanalysis

Chapter 12 deals with private theories in science. These are the private theories of scientists that escape the awareness of the scientific public. The importance of this topic for the psychoanalysis of science lies in the dynamic understanding of the individual scientist facing the scientific community. Psychological reasons that prevent the scientist from expressing his scientific opinions on certain matters are brought to the fore, given that this phenomenon is conspicuous in the field of psychoanalysis.

Chapter 13 takes up the issue of the developmental lines in the history of science, especially in psychoanalysis. Is science developing in leaps or in an evolutionary fashion?

12

Private Theories in Science

Private theories in psychoanalysis are by-products of the split psycho-analytic epistemology. Joseph Sandler defined private theories as the variety of theoretical segments that exist in analysts' preconscious in connection with their clinical work.[1] These theories remain in the preconscious, not entering the conscious as long as they do not fit the 'official' or 'public' theories. They are creative products of the analyst that remain unavailable to the knowledge of the scientific community. According to Sandler, the fact that they remain within the limits of the preconscious indicates the enormous fear of individual analysts in confronting existing 'truth'.[2]

Sandler and some of his colleagues investigated private theories related to the term 'trauma' and reached the conclusion that analysts refrain from recognizing their views to keep from being considered stupid or from being accused of heresy. Private theories are formed during the psychoanalytic dialogue between analyst and patient, and different theories thus formed may contradict one another as well as received psychoanalytical theory.

In my view, private theories reflect the wish of analysts to conduct analysis in accordance with their own insights and independently of 'public theory'. They keep these theories to themselves because they feel guilt and shame that their views may deviate from the scientific consensus. It appears to me that private theories might be conscious as well, but conscious or unconscious, they remain hidden from the world.

There are two reasons why private theories are so common in psycho-analysis. One is the field's conceptual and theoretical rigidity, and the constant tension between the need to accept its discipline and the urge to work according to individual tendencies. The enormous differences between individuals, the infinite number of combinations and possibil-ities, the diversity of emotional reactions, interactions and ways of working problems through all combine to make psychoanalysis the science *par excellence* for different approaches to judging the facts and

< PRIVATE THEORIES IN SCIENCE >

evaluating their contents and implications. The second reason for the development of private theories in psychoanalysis is connected with the first, and with the epistemological double split, which constantly confronts clinicians with discrepancies between what ought to be and their own perceptions and interpretations. In this situation it takes a lot of courage to be able to say that one has found something new or that there must be something wrong with the conventional theory.

The phenomenon of private theories exists in every science, at all times. In fact, every received public theory originates in a private theory that also must circumnavigate the rocky path from the preconscious to the conscious mind of the scientists, to consciousness and acceptance by the scientific community. The greater the tension between private and public theories, the greater the fear of the creative scientist that he or she will be criticized by peers. This results in private theories remaining clandestine. And the more this is the case, the larger the phenomenon of lacunae. In this way a great deal of the body of knowledge remains buried or completely lost.

Private theories are known from the early development of children. A classical example is the private sexual theories that children develop in addition to the 'public sexual theory' they are imbued with by their parents. Even after children are informed about sexual reality, and are said to understand it, they do not relinquish their private theories of sex for some time. This is quite a normal phenomenon.

Another example of private theories that involves children is the relationship between neologism and creativity. Children use a private language when they take their first steps in the world of speech, then they adapt themselves to the language of adults, mixing this with their private language. Eventually, they give up their private language entirely. While this is considered quite normal, one may ask whether it is a desirable development. Hyper-adaptability to adult language some-times indicates a hyper-conventional way of thinking that results in an undesirably low level of self-creativity. We are dealing, then, with an imitator. On the other hand, children who allow themselves (or are allowed) to continue using some of their private language past the time that this is considered 'normal', far from being feeble-minded or autistic, may instead be expressing the bud of authenticity and free-thinking. There is often a thin line between originality and pathological neologism. What seems indicative of deep disturbance and lack of communicative ability in one context or society might be viewed as original in a different context or society. The official, rigid attitude toward psychoanalytic theory inhibits private language and maybe even the formation of private theories.

In the scientific world, as in the private one, it is not always clear—

< EVOLUTION, REVELATION AND CREATIVITY IN PSYCHOANALYSIS >

except in retrospect—what should be viewed as neologistic thinking and what should be judged original. Whether private theories are held pre-consciously or consciously, the sad fact is that their privacy means they will most likely never reach the world. Those who are conscious of their private theories and opt to keep them private feel like children who are ashamed of their private language. Sometimes the analyst knows from his experience how ridiculous he looks in front of the public theory. Sandler believes that "the more access we can give to the preconscious theories of influenced analysts, the better we can help the advancement of psychoanalytic theory".[3] This indicates that he recognizes that much potentially useful knowledge gets lost. Although he is aware of the disease in psychoanalytic discipline, his suggestion as to the treatment of the malady is circuitous. Instead of trying to get to the bottom of it and see how we can change our science so that it can support open communication, he wants to know more about those who suffer from it.

The atmosphere in which private theories remain hidden creates a science of Marranos, in which what actually happens is concealed from public view out of fear of persecution. The lack of communication, discussion and scientific control or feedback stands in contradiction to what Popper defines as an open society.[4] A discrepancy between public and private theories invites the formation of lacunae. The greater the gap between them, the greater is the tendency for the formation of lacunae. This danger is especially great in psychoanalysis for reasons inherent to the discipline, as well as due to its developmental problems. The existence of private theories points to a symptom in psychoanalysis that must be dealt with if our science is to grow over time.

< Private Theories in Science >

Leaps versus Gradual Development

13

In this chapter we will look at the evolutionary and revolutionary aspects in the history of science and in the history of psychoanalysis. Do revolutions exist? Is our judgement in this matter an objective one or does it depend on the way we look at the scientific processes?

Clearly, the question of revolution versus evolution, or leaps versus gradual development, is complex. In some instances the answer is either/or, while in others it is both/and.

If we assume that psychoanalysis is a scientific revolution, then it stands to reason that it is part of a larger cultural revolution that began before it.

I began this book by mentioning thinkers who worked with the underground streams of science and history—Gaston Bachelard and Thomas Kuhn (philosophers of science), Claude Lévi-Strauss (an anthropologist) and Michel Foucault (a historian). Each attributes to the unconscious a function in history (including the history of science), and all agree that historical development is revolutionary rather than evolutionary. The dilemma of whether history develops by leaps or gradually is a central question in the history and philosophy of science, and therefore also in the psychoanalysis of science. This dilemma exists to the same degree with regard to socio-political history. The analogy leans on the assumption that science is undertaken by the community of scientists, which is subject to sociological and historical rules like any other social community. In what follows, we will examine the question of evolutionary versus revolutionary development with specific regard to psychoanalysis, which is a scientific discipline but also, a socio-political factor in the twenty-first century.

The analogy between a political revolution and a scientific revolution we find for the first time in Marcellin Berthelot's *The Chemical Revolution*: *Lavoisier.* Lavoisier was a chemist and a politician. In his book, Berthelot compares Levoisier's chemical revolution with the spirit of the French

< EVOLUTION, REVELATION AND CREATIVITY IN PSYCHOANALYSIS >

Revolution, both of which took shape quite suddenly in the space of fifteen years.[1]

The concept 'revolution'—the word itself is derived from *revolvere*—implies a circular movement, one which returns to a position that has been held in the past. Etymologically, the word *discover* means to cancel concealment. Discovery, as opposed to invention, does not bring something new into the world, rather, it reveals what was already there. Nevertheless, in time this concept (revolution) came to be used to indicate radical change. The etymological analysis of the word already hints at our dilemma: to what extent are revolutions in fact revolutions, that is, to what degree are they 'revolutionary'? Philosophers and historians of science are divided over this question. Some are of the opinion that in fact revolutions take place, while others believe that the development of science is gradual, even if at times it seems that the changes are sudden and drastic. The thinkers we cited in chapter 1 consider the history of science in terms of intellectual mutations, epistemological and paradigmatic crises.

We have seen that Kuhn's description of the revolutionary process is psychological. He speaks of the pre-revolutionary period in terms of scientific uncertainty (and insecurity), and he notes the tendency to deny apparent phenomenon as a result of the difficulty of accepting that existing scientific principles no longer correspond to the facts. A revolution involves a Gestaltian change in the way of one's thinking. After the revolution, people see the world in a different light. Before the revolution, denial exists, and when it erupts, a moment of surprise is brought about.[2] A similar characterization is presented by Norwood Hanson.[3]

It is no coincidence that the worldviews of Bachelard, Kuhn, Lévi-Strauss, and Foucault are determined by the way they view the history of science. They do not believe in an unconditional scientific truth, and they are of the opinion that science is the product of socio-intellectual processes. Since human behaviour is not necessarily continuous, and since thinking—especially creative thinking—is given to leaps (mutations), and inasmuch as human beings think differently during different historical and cultural periods, it is impossible to expect the history of science to be conducted in a continuous fashion. Philosophers and historians of science who believe in scientific continuity also believe that scientific truth is not dependent on human capriciousness, or on unconscious thought.

Arthur Eddington claimed that in all revolutions new words are heard to the sounds of the old music. As much as scientific truths change, so they remain the same.[4] George Sarton is of the opinion that only an initial and superficial glance yields the appearance of leaps. The deeper analysis goes, the clearer it becomes that what at first seems like

< LEAPS VERSUS GRADUAL DEVELOPMENT >

giant steps soon dwindles and eventually disappears.[5] Pierre Duhem found continuity in the development of physics starting from the fourteenth century, which led him to suspect that the physical revolution of the seventeenth century was only an illusion.[6]

The nineteenth-century philosopher Harold Hoefding arrived at the conclusion that the absolute systematization of knowledge is untenable. No manner of thought is sufficiently complete as to take in all of reality. There is, then, no general model of being, since phenomena of rival type can always be found. This is the point around which the battle is waged between different worldviews. Continuity and discontinuity become the basic points of view of phenomena of rival type. Beneath the surface of conscious personality lies the unconscious, from which rival philosophies spring. The conflicting personalities create conflicting paradigms. There exists an ongoing oscillation between continuity and discontinuity. Idealists attribute great meaning to the inevitable discontinuity of our consciousness, whereas the realists see things in terms of continuity. This psychological model has implications for the way in which the scientist—as well as the political philosopher—shapes his spiritual work. It isn't possible to separate personality from spiritual production.[7] Hoefding himself was disposed toward discontinuity. He agreed with Immanuel Kant that the longing for continuity is part and parcel of our essential will to find causal explanations, but that in fact this longing is psychological at root and does not necessarily possess any objective validity.

To the extent that we do in fact accept the existence of revolutions, we generally see them in terms of a struggle against the scientific past, and not in terms of a disregard for it. We can identify two conflicting vectors. The first points beyond the gravitational field of the known and familiar and makes possible the discovery of new ideas. The second, or evolutionary, vector—which is conservative—feeds on the existent and facilitates a habitat for new ideas. Nothing is created *ex nihilo*. According to Kuhn, the successful man of science embodies both the iconoclast and the man of tradition.[8]

When historians and philosophers of science argue for or against scientific revolutions, they take up the method of *tertium non datur*, which is to say they remove the third factor from the equation. They adopt the position of either gradual development or revolution. There are only two options. But it is also possible to consider an additional alternative: that judgment pertaining to this subject depends on the perspective and position of the observer. It is entirely possible that the revolutionary aspect becomes apparent only when we assume its existence from the start, or when we view a given phenomenon from a particular angle. By the same token, seen from another angle, we might

< EVOLUTION, REVELATION AND CREATIVITY IN PSYCHOANALYSIS >

conclude that the phenomenon is evolutionary. Sarton's model is perhaps instructive here: a bird's-eye view of a given phenomenon might lead us to believe it is revolutionary, while a closer analysis of it might lead to the opposite conclusion. The historiographic argument around the French Revolution is a case in point. As is well known, opinions vary on the subject and range from a tendency to view the French Revolution as the greatest revolution of all time, through the belief that its influence was limited to a two-kilometre radius around La Place de la Republique, and on to the position at the other extreme, which holds that it was not in fact a revolution at all.

Jean Jaurès explained the French Revolution as a substantive revolution. It brought about the ascendancy of the masses, since the social conflicts that were a direct result of the revolution accelerated the realization of class consciousness. Anton Barnave sees the decline of Feudalism and the gradual ascendancy of modern capitalism as a major centuries-old revolution that touches on every aspect of life. In the context of this long-term revolution there are 'eruptions', that is, revolutions that derive from the gap that opens between political power and economic power, when conflicting interests make peaceful equilibrium impossible.

The position which holds that in fact there was no French Revolution can perhaps best be seen in Alexis de Tocqueville's commenting that the political and administrative structure of France in the eighteenth century involved remains of medieval institutions. The first revolutionary construction was erected on the foundations of the old one. Equality before the law was part of the monarchy's overall programme against Feudalism. The revolution was therefore the immediate and violent conclusion of a mission into which the work of scores of generations had been invested. Albert Sorel, de Tocqueville's successor, tried to prove that the revolution accelerated tendencies that had already existed in the country's development. Beyond this, rather than explaining the revolution as having undermined the old regime, it can be seen as a natural by-product of French and European history—in other words, as an integral part of the central stream of European development.[9]

Examples of this sort are not hard to come by, and the principle is clear. In each case the ideological burden of the historiographer influenced his interpretation of history. The Marxist is inclined to see social changes that alter the balance of class power as constituting a revolution. A conservative who is put off by the idea of revolution will tend *a priori* to understand a process of this sort as evolutionary, in order to show that it would have taken place in any event, and that there hadn't been any need for a 'revolution'. Or worse, that it was detrimental to the

< Leaps versus Gradual Development >

cause for which it ostensibly struggled. Among those who believe in the existence of revolution, some see it as an ongoing process and some see it as a process that takes shape within a very limited period of time. Some combine the two models.

Identical models can be created with regard to the development of science, and we can see how such models might be applied within the framework of psychoanalysis. We would ask two basic questions: Was the appearance of psychoanalysis on the stage of science at the beginning of the twentieth century revolutionary or evolutionary? And was its development over the course of the twentieth century evolutionary or revolutionary?

These questions are not much discussed in psychoanalytic literature. Freud's own writings offer us a sense of what his own position was in this debate. It is possible to distinguish conflicting positions in his work, where we find both an acknowledgement of revolutionary scientific elements and a tendency to see scientific processes from an evolutionary point of view. Freud was influenced deeply by scientists and philosophers who thought in evolutionary terms, such as Charles Darwin, Ernst Haeckel, and Auguste Comte. He frequently repeated that every important scientific discovery has its source in the past. The scientist must ask himself what the source of his innovative idea actually is. Freud was conscious of the fact that Nicolaus Copernicus was not the first person to discover that the earth is smaller than the sun and revolves around it. Aristarchus of Samos had already determined this in the third century BCE[10] Freud mentions him in the context of his late theory of drives, which itself is born of the spirit of Empedocles, who wrote of the life-drive and death-drive as the two basic forces in man.[11] Freud writes that the philosopher Arthur Schopenhauer was the one who forced humankind to see the extent to which drives determine one's goals and ambitions.[12] As for the discovery of the interpretation of dreams, Freud himself admits that several of the ideas in his book had long been known. The idea of censorship in the dream was taken from Joseph Popper-Linkeus, for example.[13] Ludwig Berne wrote of something else we consider classically 'Freudian' in his book *How to Become a Writer in Three Days,* which proposes a system of creative writing based on the idea of free association.[14] (Freud received this book as a Bar Mitzvah present, and it clearly exerted a profound influence on him.)

Particular significance should be attributed to the fact that Freud raised such forward-looking ideas from the field of dream interpretation, (ideas of his that were anticipated by others), for this was one of his most original books. In *The Future of an Illusion* he says that the vicissitudes in scientific ideas are of a progressive and not a revolutionary nature.[15] And in "The Question of a *Weltanschauung*" he expresses a

< EVOLUTION, REVELATION AND CREATIVITY IN PSYCHOANALYSIS >

belief in the end of the history of science. At least in the more mature sciences there is even today, he says, a solid groundwork which is only modified and improved but no longer demolished.[16] On the other hand, one must note that the revolutionary idea fascinated Freud. As we have mentioned, he considered psychoanalysis to be one of the great revolutions of the modern age. The motto of *The Interpretation of Dreams,* taken from Virgil (by way of the revolutionary Immanuel Lassal), *acheronto movebo,* "If I cannot bend the powers above, then I will move those of the underworld",[17] constitutes unmistakable evidence of the Freudian revolution. In the finest tradition of scientific writing, the extensive bibliographical preface to this work offers a detailed description of the scientific theories of the dream and dreaming that preceded Freud. Still, one can't help but note that Freud took great pains to present us with this tiresome description not only in the name of scientific propriety (descriptions of this sort do not generally appear in his other works), but in order to stress the degree to which his discovery involved a scientific leap, and to make clear that it had nothing at all in common with work that preceded it.

If we proceed from the assumption that psychoanalysis is a scientific revolution, then it stands to reason that it is part of a larger cultural revolution that began before it. Psychoanalysis expresses the spirit of the times—in this case the middle to the end of the nineteenth century. According to Greek mythology, *Tiresias* promised Narcissus a long life, in exchange for his not coming to know himself. Narcissus, however, disobeyed Tiresias, looked at his reflection on the surface of the river, and disappeared into its waters. This Hellenic legend hints to us of man's fear of overly deep self-reflection. The history of culture, and particularly of the western world, teach us what the Greek legend managed to convey in a brief story. It is somewhat astonishing that the sciences turned to deal with man only over the course of the past two hundred years. The initial developmental stages of man's interest in the sciences saw him concentrating on the inanimate and the distant.[18] The ancient Greeks and Egyptians were first and foremost interested in astronomy. The fear of objects decreased to the extent that they were far away and inanimate. Apparently for this reason the heavenly bodies symbolized stability for the Greeks, while the earth was chaotic.[19] The past two hundred years, says Foucault, mark the search for the hidden internal dynamic—the obscured internal principle that might be able to explain what was apparent to the eye. Man's scientific interest goes hand in hand with the hidden internal dynamic.[20] However, the search for the hidden developed not only in the context of science. In the twentieth century revolutionary changes took place in the natural sciences as well. Gerald Holton claims that there came into

< LEAPS VERSUS GRADUAL DEVELOPMENT >

existence a clear tendency toward more aggressive study of the essence of things (the sub-cellular level in biology, the sub-atomic level in physics).[21]

Psychoanalysis is part of the same modern revolution that contributed to the observation of deeper and more enigmatic layers of reality in the various sciences, as well as in literature and art. Literature had preceded psychoanalysis (though it didn't call it by that name) by the middle of the nineteenth century: Dostoevsky's work is saturated with it. In this sense Freud was also anticipated by Henrik Ibsen, James Joyce and Arthur Schnitzler. A well-known anecdote about the latter has Freud sending him a letter of congratulation on his fiftieth birthday in which he admitted that he hadn't read any of his work so as not to be able to confirm that Schnitzler had anticipated him.[22] In the field of philosophy Friedrich Nietzsche clearly anticipated Freud as well. Nietzsche saw himself as a psychologist of culture. And in fact there is a good deal in common between the thought of the two men. The aim of Nietzschean psychology, like that of Freudian psychoanalysis, was to reveal and release a suppressed power; in both, doing so requires an overcoming of obstructive and destructive elements. The central question in Nietzsche's *The Birth of Tragedy* is "What is Dionysian?" In other words, what is a drive and how is culture shaped by its vicissitudes.[23] Freud's central project also saw human culture in terms of drives and their incarnations. Such concepts as the unconscious, repression, sublimation, the id, the super-ego, secondary gain from illness and secondary processes exist in Nietzsche as well, although his terminology differs from that of Freud. Also in connection with Nietzsche, we find Freud claiming—as with Schnitzler—that he hadn't read him for fear of influence.

Psychoanalysis is not inclined to highlight the cultural influence upon it; on the contrary, it is far more inclined to demonstrate its influence on culture. Nonetheless, in order for the revolution to be absorbed, it isn't enough for a few great minds to have conceived it. One of the arguments against the existence of scientific revolutions is that overly revolutionary phenomena are not absorbed in the consciousness of the scientific community. The very fact that a dialogue takes shape around a new idea proves that the scientific community is, to a certain extent, mature enough to cope with it. Therefore it is also reasonable to assume that if scientist A hadn't discovered it, scientist B would have eventually appeared on the stage of scientific history with precisely the same discovery. Had Freud come out with his theory at the end of the eighteenth century instead of the end of the nineteenth century, it stands to reason that it would not have been absorbed. It is highly unlikely that he would have been able to sell even a single edition of his writings.

< EVOLUTION, REVELATION AND CREATIVITY IN PSYCHOANALYSIS >

Freud was wont to complain that his theory didn't receive enough attention, but in reality something very different happened. Though he had many opponents, principally in Vienna ("A prophet is not without honour, save in his own country"), his theory spread like wildfire. The twentieth century, it turns out, was ripe to absorb his theory, and it struck root. True, psychoanalysis was attacked from the start, beginning with the poisonous sarcasm of the Viennese Association of Physicians and running through the writings of Europe's greatest satirist, Karl Kraus—who claimed that psychoanalysis was itself the illness it sought to cure—and on to the methodological critique of Adolph Gruenbaum in the eighties.[24] Despite it all, psychoanalysis became a twentieth-century fixture.

Freud noted the three great revolutions of the modern era: the Copernican, the Darwinian, and the psychoanalytic. (He omitted the first and second physical revolutions of the twentieth century for some reason, and he did not live to experience the genetic revolution.) The Copernican revolution clarified for humankind that the earth is not the centre of the cosmos. The Darwinian revolution revealed to man that he is merely another creature in a world of creatures. The psychoanalytic revolution also made it clear that man was not the master of his own home.[25]

The Copernican revolution in no way changed anything about life on earth. For the vast majority of its inhabitants, it made not the slightest difference whether the sun revolved around the earth or vice versa. Copernicus's discovery was, however, a momentous event for physics, philosophers and theologians. Beyond the biological dimension, Darwinism influenced social thought as well and gave rise to a religious controversy. But in the twentieth century Darwinism found itself in the shadow of the genetic revolution. It would seem that the psychoanalytic revolution trickled down into the culture at large more fully than its predecessors. It touched the personal lives of people in the twentieth century and changed the self-image of humanity. Likewise it influenced the other sciences and the arts of the twentieth century. In fact, there is virtually not a single subject in the world with which it isn't concerned.

What are the specific changes brought about by psychoanalysis that justify its being called a revolution?

The first is the idea that sexuality in childhood is part of human development which did not exist prior to Freud. Freud widened the concept of sexuality to other areas of the body, beyond the sexual organs, and he established that the juvenile sexual activity does not satisfy biological needs alone. Psychoanalysis discovered sexuality's influence over the development of the ego on the emotional and cognitive level.

< Leaps versus Gradual Development >

The addition of an unconscious element to *homo sapiens* in the twentieth century complicated considerably the question of man's responsibility for his actions. Is man a deterministic creature who has no control over what he does? This question becomes part of the twentieth-century *Zeitgeist*.

Psychiatry of the nineteenth century established clear borders between psychopathology and emotional health. Freud, on the other hand, describes a continuum between psychopathology and health. The concept of normality does not figure in his writing. In this context we have mentioned sexual deviation. In the field of neurosis we can find a similar phenomenon. That is, there is no clear distinction between psychopathology of neurosis and normality. The psychopathology of daily life is a perfect example of this: dreams, slips of the tongue, jokes.[26] Freud describes neurotic phenomena that appear among normal people—phenomena that involve phobias and compulsiveness. That these are classified as pathological is more a question of quantity than quality. Freud, in other word, pioneered a new tradition that blurred the borders of psychopathology.

Continuing the list of changes brought about by psychoanalysis, we must mention universalization and individuation. Psychoanalysis presents two conflicting tendencies in its description of people: the universalization of the human being and his individuation. With regard to universalization, Freud was hardly an innovator, and he followed in the footsteps of philosophers and anthropologists who had written in this spirit. His theory is, in this sense, ahistorical. Basic psychological principles exist that explain the development of mankind, and these principles are universal. In *Totem and Taboo*, Freud shows us how guilt feelings developed in the wake of the murder of the primal father. Freud argued that this was the basic mechanism at work in human culture. The brothers' union in guilt, following the murder of the father, is the cornerstone of social development.[27] All of human culture is constructed by Freud on a few fears and complexes, in combination with several defences against these fears and complexes. That said, clinical psychoanalysis showed, for the first time, just how different people are from one another, despite these common universal motives. The variants are infinite. Psychoanalysis revealed on the emotional level what fingerprints reveal on the biological level: no two people in the world are identical. There are many people who can be classified in the same clinical category: phobic neurosis, compulsive neurosis, hysterical neurosis, etc., but nonetheless, each of these people has a specific neurosis of his own. Each has the particular circumstances in which the neurosis took shape, and each has its own operational mechanism. Each person has his own particular neurotic

< EVOLUTION, REVELATION AND CREATIVITY IN PSYCHOANALYSIS >

style. Each brings with him the elements of his specific personality, his manner of expression and his means of enacting his defence mechanisms, as well as his personal symbolism.

Analyst Stefan Mitchel has no doubt that over the course of the twentieth century psychoanalysis experienced major changes in the field of clinical theory and meta-theory. In the clinical-theoretical field the change was from seeing man as a creature that regulates drives to understanding him as one that creates meaning. Should we see these changes as revolutionary or evolutionary? In his opinion, the question involves a matter of emphasis; it has no single, absolute answer. That said, he makes it clear that his own personal choice is to see the changes as revolutionary. He describes the motives of the evolutionists in psychoanalysis, whose central concern, as he sees it, is the preservation of Freudian theory as the heart of psychoanalysis. Developments in the field of theory are in this context to be seen merely as additions and adaptations. They must be compatible with the original theories, or at any rate not be incompatible with them. According to the evolutionists, there is nothing that cannot be explained in the context of the original theory, since Freud's writings are so rich and multi-layered, and contain everything. Those who claim that new theories are incompatible with the original theory simply have not understood the Freudian text, or have offered improper interpretations of it. There is no need to choose between different theories, since all exist under a single hat. The revolutionaries suffer from 'paradigmatic arrogance', which leads to useless paradigmatic competition. They invest their thought in narcissistic identification with perfection. Those who hold that traditional psychoanalytic thought is insufficiently rational and coherent are themselves not rational.[28]

Psychoanalytic interpretation in the context of the therapeutic process offers a helpful illustration of the dilemma. Interpretation that leads to insight allows for a new kind of observation and experience on the emotional and sensual level. Isn't this in fact a revolution in miniature? In many instances, however, the 'revolution' has been preceded by extensive preparatory work, in which memories, images and emotions have repeatedly surfaced, only to be linked through therapeutic intervention. Superficially no results of this emotional work are apparent over time, but it stands to reason that the cumulative effect of this work eventually brought about the insight at a given moment. Seen in this light, the therapeutic process appears to be evolutionary.

When water spills over a glass's rim, a drastic change takes place in the water's physical condition. It is possible, however, to describe the situation as one in which drops of water were added to the glass one after another, so that what seemed to be a drastic change brought about

< LEAPS VERSUS GRADUAL DEVELOPMENT >

by the final drop was in fact entirely dependent on the earlier drops that preceded it.

The interpretation that determines the change can be considered as revolutionary on the phenomenological level and at the same time evolutionary on the dynamic level. The question of revolution is, in this instance, a matter of perspective.

Some writers argue that Melanie Klein was a revolutionary in the field of psychoanalysis. She minimized the importance of the Oedipus complex, which had been one of the building blocks of Freudian theory, and likewise minimized the importance of the super-ego in the development of the child. She built a new psychoanalysis at the centre of which were conflicts that developed at the stage of nursing. She strengthened the image of the mother at the expense of the image of the father. She built a theory of development in which processes occur at a much earlier stage than they do in original psychoanalytic theory. What is this if not a revolution?

By the same token, however, it is possible to claim that Melanie Klein was a faithful student of Freud who filled in the missing portions of Freud's theory of development. She was faithful to the spirit of her teacher in that she emphasized the world of childhood and lent additional value to the clinical theory of the inner world. In doing so, she was closer to Freud than was his daughter Anna, who widened the scope of theory to external reality (ego psychology). Though Klein's theory aroused much greater resistance than did that of Donald Winnicott, for example, she was in fact closer to Freudian theory in her worldview than Winnicott was. Winnicott more or less abandoned the theory of drives, whereas Klein extended it. The issue of whether she was revolutionary or not in the end becomes a question for sociology, and in this instance 'revolutionary' is considered a pejorative. It is possible to demonstrate this by means of the argument between Freudians and Kleinians over the death-drive. The question under discussion was whether this idea derived from clinical conclusions or if it in fact went beyond the realm of psychoanalysis. As we know, most of the Freudians opposed the idea of the existence of the death-drive, even though Freud himself had conceived of it. This was one of the rare instances during this period when Freudians openly opposed an idea held by the founding father of the science. But it is not difficult to explain their opposition. Freud himself wrote that the idea was speculative. In doing so he gave tacit approval from the start to the resistance the idea would arouse. The claim at the time was that Freud's idea wasn't grounded in inductive clinical observation. Melanie Klein developed the idea in a clinical direction. In their debate with the Kleinians, the Freudians argued that Freud did not in any way intend to alter the theory of drives on which the

< EVOLUTION, REVELATION AND CREATIVITY IN PSYCHOANALYSIS >

psychoanalytic clinic is based; rather, he sought to explain the phenomenon of masochism, for which the theory of drives was unable to account. In seeking to explain such phenomena as autoeroticism and narcissism by means of the death-drive, Klein went beyond the limits of theory. She turned a small, systematic and scientific step—suddenly Freud's idea was no longer speculative—into a major speculative leap and thereby went beyond the borders of psychoanalytic science.[29] The Freudians were the keepers of the Freudian flame. From the Kleinian point of view, it was precisely the opposite: Klein was the one who adopted the important Freudian idea, at a time when the Freudians themselves rejected it.

The hidden but actual question being discussed at that conference was: Who is Freud's legitimate daughter and heir, Anna Freud or Melanie Klein? In the context of this debate, 'revolutionary' was a negative characterization—even though all the participants had expressed such pride in psychoanalysis as a revolutionary science. In the instance before us, however, the central matter was to demonstrate that the other party involved had betrayed Freud. This is perhaps what Kuhn had in mind when he spoke of the behaviour of the scientific community. What matters in science is the sociological principles of the scientific community; what is normative and what is exceptional are established according to the principles the community holds. What one side characterizes as evolutionary, the other will characterize as revolutionary.

Another example of conflicting interpretations of scientific development (evolutionary or revolutionary) is the issue of countertransference in clinical psychoanalytic theory. This development was discussed in chapter 10. One can see in the work of Devereux and Schafer turning points in thought relating to countertransference and in the psychoanalytic theory of cognition—thought that altered psychoanalysis in the direction of the post-modernist worldview. It is also possible to argue that all the steps that preceded those taken by Devereux and Schafer were small steps leading in evolutionary fashion to the inevitable conclusion arrived at by the two men—and that without these earlier steps, the new theory would not have been absorbed.

In addition to the examples offered here, in which the revolutionary dimension is a matter of perspective, it is possible to offer examples in which there appears to be a substantive leap in scientific thought. When we examine the Freudian worldview in terms of the etiology of emotional disturbance, we can distinguish leaps of this sort. Initially, Freud held a theory of trauma. The reasons for the neuroses he treated were, according to his diagnosis, sexual trauma, usually in childhood. In time he came to conclude that it was more reasonable that the explanation lay hidden in childhood fantasy. In other words, Freud

< LEAPS VERSUS GRADUAL DEVELOPMENT >

exchanged an etiology linked to external reality for one linked to the internal and emotional reality of the child. Such an alteration fundamentally changes clinical observation and would have profound implications for clinical conclusions drawn from this observation, regardless of the onlooker's perspective. If Freud had continued to maintain his trauma theory, psychoanalysis would have gone in another direction entirely, since its central concern was to clarify reality as a pathogenic factor and not to develop the unconscious, interior and emotional world. This is the case also with the passage from the initial drive theory to the theory based on the balance between Eros and Thanatos, which can be traced back to 1920. Clinical psychoanalytic interpretation would change in the wake of this turning point. If up until then every phenomenon was interpreted in accordance with the pleasure principle, from then on at least part of the clinical psychoanalytic interpretation would be based on the death-drive. The theoretical turning point altered the view of social psychoanalysis as well.

Clearly the question of revolution versus evolution, or leaps in science versus gradual development, is complex. In some instances the answer is either/or, while in others it is both/and.

The revolutions that took place in psychoanalysis over time were usually 'white', that is, they took place in a kind of underground fashion. Only rarely does one find in the psychoanalytic literature a direct confrontation with Freud. When Donald Winnicott exchanged the Freudian theory of drives for his own clinical object-relations theory (which related to emotional connections rather than drives), he never directly stated that the theory of drives seemed to him clinically inappropriate.

< EVOLUTION, REVELATION AND CREATIVITY IN PSYCHOANALYSIS >

Epilogue: *Scientific Creativity in the Shadows*

The aim of this book is to demonstrate the existence of irrational elements in science and to indicate how scientific use can made of them. In the introductory chapters, I gave some examples from various scientific disciplines in order to demonstrate that we are dealing with a universal phenomenon. The main part of the book, however, was dedicated to the task of showing how to work with the method described in Part I in the field of psychoanalysis. I have tried to show that the conception of irrationality in science is not based exclusively on either philosophical assumptions or Goedel's mathematical theorem. In other words, I claim this phenomenon can be tested on every science individually. If it can be successfully demonstrated that unconscious lacunae can and have affected scientific matters, this will prove that irrationality in science does not stem from the subjectivity of the individual scientist alone, but is endemic to science *per se*.

A second major question concerns what I mean by science, and whether for the sake of this work a distinction should be made between natural sciences and hermeneutic sciences. In chapter 1 I suggest that the hermeneutical exploration of scientific texts should be the basis for the psychoanalysis of science. But this does not exclude non-hermeneutical sciences, because I believe that the strict dichotomy between these two categories is somewhat artificial. The assumption that there is irrationality in science does not distinguish between different fields of science. After all, the subjective motives uncovered by Kuhn referred to the natural sciences. Every science is based on texts. While a hermeneutic science obviously provides more metaphors and scientific parapraxes, contradictions and paradoxes can and have been found in natural science texts as well. Both types of science are also equally prone to scientific myths.

The history and philosophy of science in general ask questions about the general development of science. There is no reason why these ideas cannot be transposed to research on the developmental aspects of the

< SCIENTIFIC CREATIVITY IN THE SHADOWS >

individual sciences. Taking psychoanalysis as our specific field of science can be especially useful in the investigation of hidden motives in the sciences. Using it, I have engaged in a hermeneutic investigation of four elements that serve as guidelines: metaphors, scientific parapraxes, lacunae and myths (mythologies). The metaphors lend the scientific text a personal flavour and therefore introduce us into the subjective vision of the scientist. The scientific parapraxes consist of the contradictions, paradoxes, slips, truisms and other irregularities that reveal the scientific unconscious of the scientist. Myths are the narrative aspects of science. Lacunae are the blind spots in science—those areas of science which were overlooked for unconscious reasons.

The key principle of the structuralistic method is the assumption that there is a tacit, hidden structure ordering every system. Foucault contends that the most important characteristics are those that are the most hidden. According to him we are expected to see the connections between the 'superficial' phenomena via the basic functions and motives that we discover underneath. Psychoanalysis is a major tool in the uncovering of such hidden structures. I propose it, as well, as the object of such research because it is my working hypothesis that many of the developmental aspects of psychoanalysis are tacit and the field would benefit by their being uncovered.

My book cannot be considered a systematic study that covers all the unconscious inhibitions within the psychoanalytic discipline. It is, rather, an experiment aimed at finding out how and to what extent one can apply psychoanalysis to the study of science. Obviously there are numerous blind spots in psychoanalysis with which I did not deal. In the post-Freudian psychoanalytic literature, for instance, an enormous weight is given to the function of the mother in the emotional development of the child (especially in the pre-Oedipal period). Fatherhood seems rather to be a rudimentary phenomenon. This disequilibrium in the analysis of parental functioning appears to me to be worthy of research, especially as the emotional involvement of the father seems in our times so important from the outset.

A second example of a major lacuna with which I did not concern myself is the psychoanalysis of the aged. The psychoanalytic body of knowledge in this area is very limited, as psychoanalysts are inclined to see old and elder people as poor candidates for psychoanalytic treatment. Why is this so? Those who have analysed old people know from their experience how wrong it is to disregard them *a priori* from this sort of treatment.

The delayed development of the concept of aggression in psychoanalysis constitutes a third example, and there are, of course, many others.

< Epilogue >

Generally speaking, the developmental line in psychoanalysis can be broken down into three phases: (1) there are changes in theory based on clinical experience; (2) stagnation or arrest in aspects of the discipline; (3) and a splitting of the discipline into different schools. Although I have dealt almost exclusively with stagnation, a few remarks are in order about the other two categories.

The main focus of this book has been stagnation in the scientific development of psychoanalysis. It is my thesis that this stagnation is most often due to the existence of blind spots and inhibitions in finding out about certain subjects—subjects that include the consequences of this stagnation or inhibited development. While I have utilized scientific parapraxes and metaphors routinely as a means of exploring this open question, I discussed the lacunae in a conscious and focused way. As lacunae are expressions of scientific symptoms, they must be identified in order to be worked with. The lacunae in question are the unconscious ones that conceal repressed motives and lead to scientific stagnation.

Among the symptoms of stagnation that come up in this work is the phenomenon of private theories. The greater the split between private and public theories, the more the wheels of scientific creativity are inhibited. This rule holds for all sciences, but it plays a special role in psychoanalysis.

A second major symptom is what Kuhn has called 'the distress of perception' or obstacles to bridging the gap between what one actually thinks and what thoughts are permissible in the psychoanalytic community. Kuhn is interested in scientific symptoms as they relate to paradigmatic crisis. In my work, however, scientific symptoms have not generally been the precursors to changes in paradigm, but have accompanied psychoanalysis throughout its development. Thus, they may be seen more as a chronic ailment or discontent than as a sign of dramatic change to come. The problems of traumatic anxiety and the inner hesitancy to deal with inner space have been endemic to psychoanalysis for dozens of years. This has prevented revolutionary discoveries and led to long periods of stagnation. It is not my purpose to prove or disprove the theory of paradigms or to consider scientific symptoms independent of them. In fact, I view such symptoms as an integral part of all science. Whether or not they are precursors of revolutionary changes, one must recognize them in order to evaluate their effect. For example, recognition of the epistemological split allows one to conclude that there was a crisis and that this resulted in a gradual change of approach to the psychoanalytic dyad. In this case, clinical-theoretical changes preceded meta-psychological changes.

Another scientific symptom also mentioned by Kuhn is the prolifer-

< SCIENTIFIC CREATIVITY IN THE SHADOWS >

ation of many versions of one theory. We have demonstrated this in the context of the definitions of trauma and countertransference. There is a limit beyond which one cannot claim that interpretations of a theory or a scientific term indicate scientific creativity. Beyond this limit looms stagnation.

On the other hand, hypertrophy in one aspect of a theory, such as the development of ego psychology, pushed psychoanalysis toward the outer world at the expense of the inner experiences of the self. This led to major aspects of Freud's work being neglected for a long time. As a result, the unconscious—Freud's most revolutionary discovery—suffered relative atrophy. This is Lacan's main criticism of the development of psychoanalysis.

The two major examples of Kuhn's 'distress in perception', or distur-bances in scientific perception, are unconscious lacunae and the tendency to disavow contradictions—both of which have affected psychoanalysis and science in general. That lacunae are perceptual disturbances can be seen quite clearly when blind spots in science are not recognized. Thus, the spontaneous reaction of my colleagues when I discussed the lack of psychoanalytic material on the inside of the body was "This is not possible!" or "While I have not covered the literature on the subject, I think that you are wrong here, because I had a case where . . ." These responses clearly indicate emotional reactions of surprise and disavowal that have nothing to do with scientific dialogue. The spontaneous reactions of surprise and disavowal are quite well known to analysts from sessions with patients.

While there are many theories in psychoanalysis that do not easily lend themselves to clear-cut proof or refutation, we can investigate the hypothesis that lacunae exist by checking the literature to see if anything has been written about a given theory, and, if so, what. Although one can find systematic complaints in the literature that there is lack of interest in one or another matter, none of the complainers have asked themselves what might be going on.

The theory of unconscious lacunae can be arrived at deductively as well as inductively. We can deduce their existence from the assumption that the personal defence mechanisms of scientists are at work in their scientific research, as they are in every other aspect of what they think and do. An additional assumption is that the scientific community does not necessarily correct individual blind spots, and may indeed suffer from unconscious blind spots of its own. On the other hand, we can conclude inductively that unconscious lacunae exist in science because we come across them from time to time. This was my experience when I could not understand why the traumatic dream was not considered a gold mine in psychoanalysis. When I became interested in such excep-

< EPILOGUE >

tions as this and the dearth of information on the inside of the body, I found myself in the midst of lacunae.

Systematic work on unconscious lacunae as a means of mapping and interpreting the development of science does not depend on the hermetic mapping of lacunae, or even suggest that this is necessary. Such a mapping is logically impossible because, by the very process of mapping lacunae, we produce new blind spots and thereby create new lacunae. In addition, not every unconscious lacuna is of interest for science.

When an unconscious scientific lacuna is discovered, it means that something which was repressed has come to the surface. As in psychoanalytic treatment, the coming up from the repressed causes new scientific associations. This allows one to view problems along broader lines, to detect the importance of the exceptional, to arrive at new connections with established theories. Thus, upon realizing the clinical importance of the traumatic dream, I began to search the literature for information about it. I found that it is the stepchild of dream theory— in fact, a lacuna. This led me to wonder why. And from there I began to connect the monistic dream function with the meaning of traumatic anxiety. In short, this journey led me to discover the confusion that exists around the theory of anxiety and the encapsulation of the traumatic neurosis outside the typical neurosis. Recognizing that there is a lacuna about the inside of the body led me to connect the menstrual and the incest taboos. This is how things work in science.

One's curiosity is aroused, and when one fails to find much about it in the literature, one becomes even more curious. This curiosity about the original problem, together with the interest aroused by finding that the problem represents a lacuna, opens one up to new discoveries. Thus, I will consider this book a success if it encourages others to engage in systematizing research on lacunae in science, research that views itself as a complement to our *accepted* methods of research. For there seems good reason to believe that this method will prove no less productive.

The lacunae discussed in this book represent three different types, each resulting from a different hidden motive. (1) The post-traumatic dream as a lacuna represents the mechanism of blurring a scientific problem in order to preserve the intactness of a theory. It indicates difficulty in seeing or acknowledging exceptions to a theory and results in the blockage of new developmental avenues. (2) The double split in psychoanalysis reflects the discrepancy between what we want to see and what we do see (when we see at all). (3) The lacuna about the inside of the body centres on the difficulty in seeing and explaining phenomena due to basic, perhaps inherent, fears and the anxiety they evoke.

< SCIENTIFIC CREATIVITY IN THE SHADOWS >

The motives that underlie all three of these lacunae are basic to science, and universal. Every science experiences difficulties in recognizing the exceptional and fears change. No science is free of epistemological splits. Nor is there any science in which research does not arouse some fears and anxieties.

Although my choice of these specific lacunae seems to have occurred by chance, as I never deliberately set out to deal with any of them as lacunae, it occurred to me *post-factum* that I must have stuck with these subjects because they occupied my preconscious mind.

In analysing the degree to which each of these lacunae has inhibited or completely prevented discussion of the subjects they represent, I have come up with the following categorization. The split in epistemology has to some extent been raised, especially in the clinical context. It has been an issue in psychoanalysis since the 1960s, but not all the relevant conclusions have been drawn. Trauma, traumatic anxiety and the post-traumatic dream, on the other hand, have been covered up to preserve the harmony and intactness of Freud's theory. The issue of fears surrounding the inside of the body is even more tacit because these fears represent contents that are disturbing and frightening on a more basic, almost universal level.

My interest in nightmares, especially in post-traumatic dreams, and in fantasies about the inside of the body, developed independently and resulted in different types of conclusions as to why they represent lacunae. Yet nightmares and fantasies about inner space share primary fears that may be viewed as primordial. Nightmares can be considered primordial because of the primary unmastered anxiety involved, as well as their occasional demonic content. (While the post-traumatic dream is, of course, pathological, it belongs to the family of nightmares.) The primordiality of the inner space is connected with its representation of the most basic feelings and anxieties about the womb, falling out of the womb and falling into the world.

Everybody experiences nightmares and uncontrolled anxiety, terror and the disorientation they bring upon awakening. Everybody has reflexive fears of falling in space, even if only in dreams. It seems that analysts find it easier to deal with psychopathological situations that seem remote from normality than with experiences that resonate with our own. For the latter show how frightening our inner world is to all of us, an insight that is very shocking.

As expressed in traumatization, Thanatos represents psychic death reflected in the arrest of intra-psychic activity. The post-traumatic dream is a representation of the death instinct. Thus, symbolically the traumatic dream represents the most naked experience of our fear of annihilation. It is an objectless state in which objects cease to play any

< Epilogue >

role. It is an experience of total rupture and loneliness. It is an existential trauma. The inner space also approaches basic atavistic features such as coming in touch with our naked, unprotected selves. This fear is so immense that sensations of the inner space are mostly projected to the outside. Fears of annihilation and of being dropped, which are connected with the inner space, are also objectless states.

When we turn to the subject of epistemology, we are also dealing with a basic emotional problem and not just a philosophical one. The human dilemma in respect to epistemology is the oscillation between narcissism and the oceanic urge. When a person reaches too much from himself toward outer objects, he sinks into oceanic feelings and is overwhelmed by the outside world to the point where he loses himself. On the other hand, those who lean too strongly toward the narcissistic pole lose the objects around them. This difficulty is reflected in the myth of Narcissus. Narcissus turned away from the world, but he did not learn to know himself as well. In order to understand this myth we must remember that it had its beginning in Teresias' prophecy that Narcissus would live to old age on the condition that he never look into himself. When Narcissus dared to take a deep look inside himself, he disappeared. In my view, this myth has been carried over to the human fear of knowing the essence of existence. When the individual pushes it too far, he gets into ontological confusion, but when he endeavours to fuse with the world in oceanic desire, he also disappears.

In endeavouring to find the border between themselves and the world, scientists are supposed to be able to recognize phenomena and make the right connection between them and their inner world (the process of interpreting scientific data). They lose perspective if they are overly inclined in the direction of the oceanic. If, on the other hand, they are overly inclined to hold back from the object and build a strong epistemological wall around themselves, they risk losing their creativity. Every scientist moves back and forth along the spectrum that stretches between oceanic sinking and narcissistic sinking. Most of the time we seem to be wandering somewhere between those two dangers. The art is to find the modus.

Scientific myths, like lacunae, are important cornerstones for the psychoanalysis of science.

In this book I demonstrated how Freud used myth in an unconscious way in order to maintain the integrity of his drive theory and in order to save the Logos from this theory in order to grant it an extraterritorial and independent status. Scientific myths do exist in all sciences, and their exposure may uncover some deeper motives in their development.

Another way to investigate unconscious motives in science is by exploring the rhetoric within the scientific text. I demonstrated some

< SCIENTIFIC CREATIVITY IN THE SHADOWS >

examples of Freud's rhetoric. The metaphor of the dictatorship of the Logos is an idiosyncratic and impressive expression which describes, more than any scholared text, Freud's clinging to the idea of Logos in a defensive fashion. Another example is Freud's emotional description of his attitude towards the oceanic feelings, which reveal his aversion for this sort of epistemology.

I should like to close by bringing to the light one last lacuna, the one that was the driving force behind my writing this book, namely, that psychoanalysis, which has exhibited such vivid interest in so many aspects of culture, has never used its tools to investigate any form of science (including itself). Marshall Bush had pointed out that scientific interest and creativity received surprisingly little detailed consideration in psychoanalysis.[1] The few psychoanalysts who have done work on scientific creativity, for example Peter Giovacchini,[2] Ernst Kris,[3] Lawrence Kubie,[4] have researched creativity in general, leaving scientific creativity in the shadows.

Interestingly, material on the psychology of science—in so far as it exists at all, has been left in the hands of gifted layman—the journalist Arthur Koestler[5] and the philosopher Jacques Hadamard.[6] Psychoanalysis has completely ignored science as a subject for research.

If there is an unconscious lacuna that is easy to explain, it is this neglect of science by psychoanalysis. For, although it appears on the surface that psychoanalysis would exhibit deep interest in science as one of the cornerstones of our culture and as evidence of man's psychic power, it is precisely the pivotal nature of scientific understanding that has prevented psychoanalysis from taking an interest in the subject as a source of research.

Science is the only aspect of man's creation that Freudian psychoanalysis has left outside drive theory. Freud considered science the epitome of human achievement and humanity's only hope. As everything in the domain of the drives is doomed to destruction by human ambivalence and the forces of the death instinct, science—the modern secular religion—must remain above all this. As our only hope for the future, our only solid evidence of the value of our existence, we cannot allow it to be viewed under the cruel lens of psychoanalytic investigation, for this might result in doubts that would weaken the underpinnings of our very being. Freud left this cultural testament of scientific infallibility to the psychoanalytic community, which—unconsciously—has remained faithful to it until today.

< EPILOGUE >

Notes

Part I   *Locating the Irrational, Understanding the Repressed*

Chapter 1   Why a Psychoanalysis of Science?

1   Gaston Bachelard, *The New Scientific Spirit*, trans. A. Goldhammer, Boston: Beacon Press (1984), p. 13.
2   Gaston Bachelard, *Le Matérialisme rationnel*, Paris: Presses Universitaire de France (1972), p. 50.
3   Bachelard, *The New Scientific Spirit*, p. 17.
4   Gaston Bachelard, *The Psychoanalysis of Fire*, Boston: Beacon Press (1964), pp. 21–2.
5   Claude Lévi-Strauss, *The Savage Mind*, Chicago: Chicago University Press (1966), p. 95.
6   *Ibid.*
7   Claude Lévi-Strauss, *Structural Anthropology*, trans. Clair Jacobson, New York: Basic Books (1963), p. 203.
8   Ino Rossi, *The Unconscious in Culture*, New York: E. P. Dutton & Co. Inc. (1974), pp. 7–28.
9   "An Interview with Michel Foucault", *Quinzaine Litteraire*, March 1958.
10 Michel Foucault, *The Order of Things: An Archaeology of the Human Sciences*, New York: Vintage Books (1970), pp. 3–16.
11 Jan Ehrenwald, "History of Psychoanalysis", *Science and Psychoanalysis*, ed. Jules Masserman, New York: Grune & Stratton (1958), p. 148.
12 Michel Vovell, *Ideologies and Mentalities*, trans. Eamon O'Flaherty, Cambridge: Polity Press (1990), p. 136.
13 Philipe Ariès, *The Hour of our Death*, trans. Helen Weaver, Oxford: Oxford University Press (1981), p. xvii.
14 Vovell, *Ideologies and Mentalities*, pp. 77–80.
15 *Ibid.*, p. 9.
16 *Ibid.*, p. 72.
17 *Ibid.*, p. 136.
18 Ibid, p. 136.
19 Dominique LaCapra, "History and Psychoanalysis", in F. Meltzer (ed.), *The Trials of Psychoanalysis*, Chicago: Chicago University Press (1987), pp. 9–38.
20 *Ibid.*

< NOTES TO PP. 3–8>

21 Dominique LaCapra, *Historical Criticism*, Ithaca: Cornell University Press (1985), pp. 71–94.

22 Dominique LaCapra, "Representing the Holocaust", in S. Friedlaender, *Probing the Limits of Representation: Nazism and the Final Solution*, Cambridge, MA: Harvard University Press (1992), pp. 71–94.

23 *Ibid.*, p. 111.

24 Hayden White, *The Content of the Form*, Baltimore: The Johns Hopkins University Press (1987), pp. 22–5.

25 Roy Schafer, *The Analytic Attitude*, New York: Basic Books (1983), pp. 7–8.

26 Saul Friedlaender, *Probing the Limits of Representation*, p. 6.

27 Schafer, *The Analytic Attitude*, pp. 7–8.

28 White, *The Content of the Form*, p. 1.

29 Henry Lawton, "Psychohystorical Methodology", in *The Psychohistorian's Handbook*, New York: Psychohistory Press (1988), p. 85.

30 *Ibid.*, p. 86.

31 Peter Loewenberg, *Decoding the Past*, Berkeley, CA: California University Press (1983), pp. 4–5, 17.

32 Thomas Kuhn, *The Structure of Scientific Revolutions*, Chicago: University of Chicago Press, 2nd edition (1970), p. 67. According to Kuhn the characteristics of new discoveries are: the previous awareness of anomaly; the gradual and simultaneous emergence of both observational and conceptual recognition; and the consequent change of paradigm categories often accompanied by resistance. *There is evidence that these characteristics are built into the nature of the perceptual process itself* (p. 62).

33 Ludwig Fleck, *Genesis and Development of a Scientific Fact*, Chicago: University of Chicago Press (1979), p. 21.

34 *Ibid.*, pp. 22–3.

35 *Ibid.*, p. 27.

36 *Ibid.*, p. 38.

37 Paul Feyerabend, *Against Method*, London and New York: Verso (1988), pp. 9–13.

38 Norwood Russell Hanson, *Patterns of Discovery*, Cambridge: Cambridge University Press (1958), pp. 4–34.

39 Michael Polanyi, *Personal Knowledge*, London: Routledge & Kegan Paul (1962), pp. 20, 135.

40 Peter B. Medawar, *Induction and Intuition in Scientific Thought*, Philadelphia: American Philosophical Society (1969), pp. 42–59.

41 Karl Popper, *Realism and the Aim of Science*, London: Hutchinson (1956), p. 92.

42 Medawar, *Scientific Thought*; Karl Popper, *The Logic of Scientific Discovery*, New York: Harper & Row (1968), p. 92: "I do not suggest that there is no such thing as subjective knowledge . . . There is even reason for having a theory of subjective knowledge, yet any such theory would be part of an empirical science, not part of the logic of science or epistemology. For its topic is the growth of *somebody's* knowledge."

43 Feyerabend, *Against Method*, pp. 9–13.

< NOTES TO PP. 8–13>

44   Karl Popper, *Objective Knowledge*, Oxford: Clarendon Press, (1972), pp. 106–50.

45   Feyerabend, *Against Method*, pp. 9–13.

46   Karl Popper, *Conjectures and Refutations*, London: Routledge & Kegan Paul (1974), pp. 226, 231: "We have no criterion, but are nevertheless guided by the idea of truth as a regulative principle; and that, though there are no general criteria by which we can recognise truth . . . there are criteria of progress towards the truth. The very idea of error or of doubt implies the idea of an objective truth which we may fail to reach."

47   Kuhn, *The Structure of Scientific Revolution*, pp. 66–76.

48   Amos Funkenstein, *Theology and the Scientific Imagination*, Princeton, NJ: Princeton University Press (1986).

49   Gerald Holton, *The Scientific Imagination: Case Studies*, Cambridge: Cambridge University Press (1978).

50   William W. Barthley, *The Retreat to Commitment*, New York: Alfred Knopf (1962).

51   L. Laudan, *Progress and Its Limits: Towards a Theory of Scientific Growth*, Berkeley, CA: University of California Press (1972).

52   Joseph Rouse, *Knowledge and Power: Toward a Political Philosophy of Science*, Ithaca: Cornell University Press (1987).

53   Barry Barnes, *T. S. Kuhn and Social Science*, London: Macmillan (1982), pp. 58–9.

54   Steven Shapin, *A Social History of Science*, Chicago: Chicago University Press, (1994), chap. 4.

55   Anthony Giddens, *The Consequences of Modernity*, Stanford: Stanford University Press (1989), pp. 21–7, 79–85.

56   Mary Douglas, *How Institutions Think*, New York: Syracuse University Press (1986), pp. 19, 32.

57   Giambattista Vico, *The New Science*, Ithaca: Cornell University Press (1991). See for instance pp. 127–8.

58   Joseph Schelling, "Myth", in *The Encyclopaedia of Philosophy*, London: Collier Macmillan (1967), vol. 5, p. 434–7.

59   S. Freud, "Why War: A Letter from Freud to Einstein", Standard Edition 22, London: Hogarth Press (1962), p. 211: "It may perhaps seem to you that our theories are a kind of mythology, and in the present case, not even an agreeable one. But does not every science come in the end to a kind of mythology like this?"

60   Jeffrey Moussaieff Masson (ed.), *The Complete Letters of Sigmund Freud*, Wilhelm Fliess, p. 28: "The inner perception of one's own psychical apparatus stimulates illusions of thought which are naturally projected outward. Immortality, retribution, life after death, are all reflections of our inner psyche . . . psychomythology."

61   Ernst Cassirer, *The Myth of the State*, New York: Dove Publications (1946), p. 28: "By the appearance of Freudian theory came a new conception that opened a wide horizon. Myth was no longer connected with isolated facts. It was connected with well-known phenomena which could be studied in

< Notes to pp. 13–17>

a scientific way and which were capable of empirical verification. Thus, myth became perfectly logical—almost too logical."

62 Ernst Cassirer, *The Philosophy of Symbolic Forms*, 3 vols, New Haven and London: Yale University Press (1955).

63 Hans Blumenberg, *Work on Myth*, Cambridge, MA: MIT Press (1985).

64 Lévi-Strauss, *The Savage Mind*.

65 Karl Popper, *Conjectures and Refutations*, p. 226.

66 Ernst Haeckel, *Anthropogenie oder Entwicklung der Menschen*, Leipzig: Engelmann (1874).

67 Auguste Comte, *Introduction to Positive Philosophy*, ed. Frederick Ferré, Indianapolis: Bobbs–Merrill Co. Inc. (1970), pp. 1–33.

Chapter 2   Lacunae in the Development of Science

1 Kuhn, *The Structure of Scientific Revolutions*, p. 94.

2 Bernard Barber, "Resistance by Scientists to Scientific Discovery", in B. Barber and N. Hirsch (eds), *The Sociology of Science*, New York: Free Press (1962), pp. 539–56.

3 Shmuel Samburksy, *The Laws of Heaven and Earth*, Jerusalem: Bialik Institute, (1987) (Hebrew).

4 Stephen Hawking, *A Brief History of Time*, New York: Bantam Books (1988), p. 84.

5 Barber, "Resistance to Scientific Discovery". Bernard Barber points out that there has been a relative lack of attention to scientific resistance on the part of scientists themselves, who tend to deny that the evaluation of scientific discoveries is influenced by unscientific factors. Barber hints that such a resistance might be due to the notion of the stereotype of the scientist as the open-minded man.

6 David Joravsky, *The Lysenko Affair*, Cambridge, MA: Harvard University Press (1970), pp. 231–70.

7 Barber, "Resistance to Scientific Discover".

8 Oliver Sacks, "Scotoma: Forgetting and Neglect in Science", in Robert B. Silvers (ed.), *Hidden Histories of Science*, London: Granta Books (1995), pp. 141–88.

9 Marshall Bush, "Psychoanalysis and Scientific Creativity", *Journal of the American Psychoanalytic Association* 17, 1 (1969), pp. 136–89.

10 Lévi-Strauss, *The Savage Mind*, pp. 257–8.

11 The founder of modern ethology is Konrad Lorenz.

12 It seems that totemism is universal and lives with us even in the twentieth century. A striking example of discrepancies between what we imagine animals to be like and what they are in fact can be concluded from the observations of Jane J. Lawick-Goodall on chimpanzees (*In the Shadow of Man*, Boston: Houghton Mifflin [1971]). It is not only animals that we have emotional difficulty in observing directly, but also the human being. The study of direct infant observation started only in the third decade of our century. Charles Tidd mentioned that the study of animal behaviour and the direct observation of infants and children have a great deal in common ("Symposium on Psychoanalysis and Ethology", *International Journal of*

< Notes to pp. 17–23>

Psychoanalysis 41 [1960], p. 312). It seems to me that the difficulties in observing animals and human beings have something in common, as animals take on in our imagination a lot of human attributes.

13   Sambursky, *Laws*.

14   Popper, *Objective Knowledge*, pp. 107–50. Popper claims that knowledge in the objective sense is totally independent of anybody's claim to know. It is also independent of anybody's belief or disposition to assert to act. Knowledge in the objective sense is knowledge without a knower: it is a knowledge without a knowing subject.

15   Kuhn, *The Structure of Scientific Revolution*, pp. 52-91: "Copernicus complained that in his day astronomers were so inconsistent in these [astronomical] investigations . . . that they could not even explain or observe the constant length of the seasonal year. 'With them', he continued, 'it is as though an artist were to gather the hands, feet, head and other members for his images from diverse models, each part excellently drawn, not related to a single body, and since they in no way match each other, the result would be monster rather than man.' Einstein, restricted by current usage to less florid language, wrote only: 'It was as if the ground had been pulled out from under one, with no firm foundation to be seen anywhere, upon which one could have built.' Such explicit recognitions of breakdown are extremely rare, but the effects of crisis do not entirely depend upon its conscious recognition . . . All crises begin with the blurring of a paradigm and the consequent loosening of the rules for normal research."

16   Michel Foucault, *The Order of Things: An Archaeology of the Human Sciences*, New York: Vintage Books (1970), pp. 3–16.

17   James Gleick, *Chaos: Making a New Science*, New York: Viking (1987), pp. 11–31.

Chapter 3   Psychoanalytic Historiography

1   Kurt Goedel, *On Formally Undecidable Prepositions*, New York: Basic Books (1962).

2   F. Bridgman, *The Way Things Are*, Cambridge, MA: Harvard University Press, (1959), p. 6.

3   Barthley, *The Retreat to Commitment*.

4   Joseph Agassi, *Science and Society*, Dordrecht: Reidel Publishing Co., 1981), pp. 465–76.

5   André Haynal and Ernst Falzeder, *One Hundred Years of Psychoanalysis: Contributions to the History of Psychoanalysis*, London: Karnac Books (1994).

6   Arnold Davidson, "How to Do History of Psychoanalysis: A Reading of Freud's Three Essays on the Theory of Sexuality", in François Meltzer (ed.), *The Trials of Psychoanalysis*, Chicago: University of Chicago Press (1987), pp. 39–64.

7   Judith Vita and Sándor Ferenczi, "Amalgamating with the Existing body of Knowledge", in André Haynal and Ernst Falzeder, *One Hundred Years of Psychoanalysis: Contributions to the History of Psychoanalysis*, London: Karnac Books (1995), pp. 257–63.

< Notes to pp. 24–31 >

8 Frank J. Sulloway, *Freud, Biologist of the Mind*, New York: Basic Books (19797), pp. 32–44.

9 José Brunner, Oxford: Basil Blackwell (1995), Part II, pp. 47–88.

10 Davidson, "History of Psychoanalysis".

11 Sander Gilman, *The Case of Sigmund Freud*, Baltimore: The Johns Hopkins University Press (1992), p. 172.

Part II   The Lacuna of Images

Chapter 4   The Inside of the Body

1 Yehoyakim Stein, "Some Reflections on the Inner Space and Its Contents", *Psychoanalytic Study of the Child* 43 (1988), pp. 291–304.

2 In *Totem and Taboo* Freud argued that the most ancient form of sacrifice involved an animal whose flesh and blood were shared by members of the group. The partners, or brothers, were each other's guarantors. They were united physically in such a way that when one of them was killed, the others would announce: "Our blood was shed." The soul of the sacrificed animal, which lies in its flesh and blood, was acquired by the participants through the sacrifice. This is the basis for the blood treaty.

3 George Devereux, "The Psychology of Feminine Genital Bleeding", *International Journal of Psychoanalysis* 31 (1950), pp. 237–57.

4 René Spitz, "The Primal Cavity", *Psychoanalytic Studies of the Child* 10 (1955), pp. 215–41.

5 Marion Milner, *The Hand of the Living God*, London: Hogarth Press (1969), p. 272.

6 Melanie Klein, *The Psychoanalysis of Children*, London: Hogarth Press (1932). *Idem*, "The Oedipus Complex in the Light of Early Anxieties", in *Contributions to Psychoanalysis 1921–1945*, London: Hogarth Press (1945), p. 85.

7 Klein, "The Oedipus Complex", p. 85.

8 Judith Kestenberg, "Inside and Outside—Male and Female", *Journal of the American Psychoanalytic Association* 16 (1968), pp. 457–519.

9 Devereux, "Feminine Genital Bleeding".

10 Paul Schilder, "What Do We Know about the Interior of the Body?", *International Journal of Psychoanalysis* 16 (1935), pp. 355–60.

11 Eric Erikson, "Womanhood and the Inner Space", in *Identity*, New York: W. W. Norton (1968), pp. 261–94.

12 Enid Balint, "On Being Empty of Oneself", *International Journal of Psychoanalysis* 44 (1963), pp. 470–80.

13 Owen Renik, "An Example of Disavowal Involving the Menstrual Cycle", *Psychoanalytic Quarterly* 53 (1984), pp. 523–32.

14 Mary Chadwick, *The Psychological Effects of Menstruation*, New York: Nervous and Mental Disease Publishing Co. (1932), p. 26.

15 *Ibid.*, p.15.

16 *Ibid.*, p. 11.

17 Mary Lupton, *Menstruation and Psychoanalysis*, Urbana: University of Illinois Press (1993), pp. 201–6.

< Notes to pp. 31–43>

18 Devereux, *"Feminine Genital Bleeding"*.

19 Eva Lester and Malkah Notman, "Pregnancy, Development Crisis and Object Relations", *International Journal of Psychoanalysis* 67 (1968), pp. 357–66.

20 Newell Fisher, "Multiple Induced Abortion", *Journal of the American Psychoanalytic Association* 22 (1974), pp. 395–407.

21 Devereux, "Feminine Genital Bleeding".

22 Howard Steward, "Changes of Inner Space", *International Journal of Psychoanalysis* 66 (1985), pp. 255–64.

23 Erikson, "Womanhood and the Inner Space".

24 Frieda Fromm-Reichman, *Principles of Intensive Psychotherapy*, Chicago: University of Chicago Press (1950), pp. 202–88.

25 Joan Raphael-Leff, *Psychological Processes of Childbearing*, London: Chapman & Hall (1991), p. 49.

26 Joan Raphael-Leff, *Pregnancy: The Inside Story*, London: Sheldon Press (1993), p. 40.

27 John Frosch, *The Psychotic Process*, New York: International Universities Press (1983).

28 Thomas Freeman, *A Psychoanalytic Study of Psychoses*, New York: International Universities Press (1973).

29 Harold Searls, *Collected Papers on Schizophrenia and Related Subjects*, London: Hogarth Press (1965), p. 583.

30 Herbert Rosenfeld, *Psychotic States*, New York: International Universities Press (1973).

31 Heinz Kohut, *The Analysis of the Self*, New York: International Universities Press, (1971).

32 Otto Kerneberg, *Borderline Conditions and Pathological Narcissism*, New York: Jason Aronson (1975).

33 Searls, *Schizophrenia and Related Subjects*.

34 David Rosenfeld, *The Psychotic*, London: Karnac Books (1992), pp. 179–98.

35 Jannine Chasseguet-Smirgel, "Blood and Nation", Panel on the Psychodynamics of Nationalism—Past and Present, International Congress of Psychoanalysis, Amsterdam, 1993.

36 S. Freud, *The Taboo of Virginity* (1918), Standard Edition 11 (1962), p. 193.

37 S. Freud, *Totem and Taboo* (1913), Standard Edition 13 (1962), pp. 1–161.

38 Devereux, *"Feminine Genital Bleeding"*.

39 Helene Deutsch, *The Psychology of Women*, New York: Grune & Stratton, (1944), vol. 2, pp. 149–84.

40 *Ibid.*

41 Klein, *The Psychoanalysis of Children*.

42 Stein, "Inner Space and Its Contents". Supermenstruation is the image that certain tribes have of menstruation. They consider pregnancy as a sort of menstruation, and believe that pregnancy derives from menstrual blood.

43 Devereux, "Feminine Genital Bleeding".

44 Deutsch, *The Psychology of Women*, vol. 2.

45 Claude Dagmar Daly, "Der Menstruationskomplex", *Imago* 14 (1928), pp. 11–75.

< Notes to pp. 44–48>

46   Claude Dagmar Daly, "Zu meiner Arbeiten über die weiblichen Tabuvorschriften", *Zeitschrift für Psychoanalytische Pädagogik* 5–6 (1933), pp. 225–8.

47   Sándor Ferenczi, *Final Contributions to the Problem and Method of Psychoanalysis*, London: Hogarth Press and Institute of Psychoanalysis (1955), p. 122.

48   Centrale Zeitung für psychoanalytische Pädagogik, 5/6 (1931). In these works we find mainly referrals to Freud, anthropological remarks and very little clinical impressions.

49   Ruth Lidz and Theodor Lidz, "Male Menstruation: A Ritual Alternative to the Oedipal Transition", *International Journal of Psychoanalysis* 58 (1977), pp. 17—31.

50   Claude Dagmar Daly, "Hindu Mythology und Kastreationscomplex", *Imago* 13 (1927), pp. 145–98.

51   Robert Briffault, *The Mother*, London and New York: George Allen & Unwin (1959), p. 243.

52   Jean Bertrand Pontalis, *Frontiers in Psychoanalysis*, New York: International Universities Press (1977), chap. 9.

53   George Devereux *A Study of Abortion in Primitive Societies*, New York: International Universities Press (1973), p. 28.

54   Dinora Pines, "Skin Communications, Early Skin Disorders and Their Effect on Transference and Counter-Transference", *International Journal of Psychoanalysis* 61 (1980), pp. 315–22.

55   Esther Bick, "The Experience of the Skin in Early Object-Relations", *International Journal of Psychoanalysis* 49 (1968), pp. 484–6.

Chapter 5   Incest and Menstruation Taboos

1   Marvin Harris, *The Rise of the Anthropological Theory*, London: Routledge & Kegan Paul (1968), pp. 278–9.

2   George Devereux, *Ethnopsychoanalysis*, Berkeley: University of California Press (1978), pp. 80–3. The *Elementargedanken* asserts that the psychic unity of man implies that certain ideas would arise everywhere when conditions were fairly comparable. Things which are repressed in one society are often conscious and culturally fully implemented in another society. The human mind functions pretty much the same way everywhere (psychic unity).

3   James Frazer, *The New Golden Bough*, New York: Criterion Books (1959), pp. 359–61.

4   Robin Fox, *The Red Lamp of Incest*, New York: E. P. Dutton (1980), p. 4.

5   Emile Durkheim, *Incest: The Origin and Nature of the Taboo*, Lyl: Stuart (1963).

6   Joseph Shepher, *Incest: A Biosocial View*, New York and London: Academic Press (1983), chap. 3, p. 25.

7   Freud posited his theory in *Totem and Taboo*, where he showed how the taboo developed from the primary herd. The young males conspired against the leader and killed him, after which a covenant between the brothers introduced the incest taboo. The horror of incest was transmitted

< NOTES TO PP. 48–56>

from one generation to the other. Freud's basic assumption is that male attraction to females is ever active and, if not held under control (in form of social rules), would be destructive. Freud focuses entirely on the male's position while the female's position is overlooked.

8   Bronislaw Malinowski, *Sex and Repression in Savage Society*, New York: Meridian (1927). Malinowski accepted Freud's notion of sexual attraction within the family as a universal phenomenon; still, in his eyes, the expression of the attraction is a function of the social system. Malinowski discovered that the Oedipus complex does not exist among the Trobrianders and therefore concluded that it is culture-bound and not universal. The nuclear family is the cornerstone of the social order. As incest threatens the fundamental bonds of kinship, it threatens the whole of society.

9   Brenda Seligman, "The Incest Barrier", *Journal of the Royal Anthropological Institute* 59 (1929), pp. 268-9. She combined in her work a sociological and psychological explanation. She contended that within the family there is a tendency for rivalry that is inherently human. The adoption of incest laws helps to preserve harmony over periods when rivalry might otherwise become acute. A sexual relationship between parents and children might weaken the status system of the family and destroy family stability and socialization. The sanction for the incest barrier is primarily the persistence through childhood of the infant's belief in the omnipotence of the parents, later developed into a religious attitude and consolidated by rites expressive of ancestor worship. The persisting family is a social group with such revival value that societies in which incest was the rule have died out.

10 George Peter Murdoch, *Social Structure*, New York: Macmillan (1949).

11 Talcott Parsons "The Incest Taboo in Relation to Social Structure and the Socialization of the Child", *British Journal of Sociology* 5 (1954), pp. 101–7.

12 Malinowski, *Sex and Repression*.

13 Claude Lévi-Strauss, *The Elementary Structure of Kinship*, Boston: Beacon Press (1969), p. 21.

14 Edmund Taylor, "On a Method of Investigating the Development of Institution Applied to Laws of Marriage and Descent", *Journal of the Royal Anthropological Institute* 18 (1988), pp. 267–8.

15 Eduard Westermark, *The History of Human Marriage*, London: Macmillan (1921), pp. 35–85.

16 Durkheim, *Incest*, pp. 89–90.

17 Lévi-Strauss, *The Elementary Structure of Kinship*, p. 21.

18 Briffault, *The Mother*, p. 243.

19 *Ibid.*

20 Devereux, *"Feminine Genital Bleeding"*.

21 Karen Horney, "The Dread of Women", *International Journal of Psychoanalysis* 13 (1932), pp. 348–60.

22 Durkeheim, *Incest*, pp. 87–96.

23 Alfred Gell, "Reflections on a Cut Finger", in R. H. Hook (ed.), *Phantasy and Symbol*, New York: Academic Press (1979), pp. 137–41.

24 Freud, *Totem and Taboo*.

< NOTES TO PP. 56–61 >

25   *Ibid.*
26   Erich Neumann, *The Great Mother*, Princeton: Princeton University Press (1963), pp. 71, 149–70.

Part III   *Scientific Myths and Lacunae: From* Thanatos *to* Logos

Chapter 6   The Deductive Birth of *Thanatos* Theory

1   S. Freud, *Beyond the Pleasure Principle* (1920), Standard Edition 17 (1973), pp. 7–64.
2   This is the only instance in which Freud refers to a normal activity, which leads one to suspect him of being on the defensive *vis-à-vis* his being abandoned by his best friends. It is fate, not neurotic traits, which brings about a pattern of being abandoned.
3   S. Freud, *Instincts and Their Vicissitudes* (1915), Standard Edition 14 (1973), pp. 117–40.
4   In *Beyond the Pleasure Principle* Freud claims that the death instincts are basically mute, that words basically arise from Eros.
5   S. Freud, "Revision of the Theory of Dreams" (1933), Standard Edition 22 (1973), pp. 7–30.
6   Freud, *Beyond the Pleasure Principle*, pp. 7–64.
7   *Ibid.* To the best of my knowledge, Freud never before or after expressed such subjective feelings of despair in relation to scientific matters.
8   Kuhn, *The Structure of Scientific Revolution*, p. 94.
9   Max Schur, *Freud: Living and Dying*, New York: International Universities Press, (1972), p. 326.
10   Martin Stein, "States of Consciousness in the Analytic Situation Including a Note on the Traumatic Dream", in M. Schur (ed.), *Drives, Affects and Behavior*, New York: International Universities Press (1965), pp. 61–86.
11   Theodor Lidz, "Nightmares and Combat Neurosis", *Psychiatry* 9 (1946), pp. 37–49.
12   John Mack, "Nightmares, Conflict and Ego Development in Childhood", *International Journal of Psychoanalysis* 45 (1965), pp. 403–28.
13   John Mack, "Dreams and Psychosis", *Journal of the American Psychoanalytic Association* 17 (1969), pp. 206–21.
14   Charles Fisher, "A Psychophysiological Study of Nightmares and Night Terrors", *Journal of Nervous and Mental Diseases* 2 (1973), pp. 75–97.
15   *Ibid.*
16   S. Freud, *The Unconscious* (1915), Standard Edition 14 (1973), pp. 214–15.
17   S. Freud, *The Interpretation of Dreams* (1900), Standard Edition 4–5 (1973), pp. 124–5, 130, 161.
18   Freud gives an example of a man who ate salt fish in the evening and dreamt that he was drinking cold water. In Freud's interpretation, the only function of this dream was to extend sleeping; the content had no symbolic meaning.
19   Laurence Kubie, "A Reconsideration of Thinking the Dream Process and the Dreams", *Psychoanalytic Quarterly* 35 (1966), pp. 191–8.

< Notes to pp. 61–74>

20   Laurence Kubie, "A Critical Analysis of the Concept of Repetition Compulsion", *International Journal of Psychoanalysis* 20, 3/4 (1939), pp. 390–402.

21   Leon Saul, "Psychological Factors in Combat Fatigue", *Psychosomatic Medicine* 7, 5 (1945), pp. 257–72. Franz Alexander preceded Saul in his presenting the notion that "in using the concept of repression . . . we never deal with a simple photographic repetition of a former pattern" (*Psychoanalysis and Psychotherapy*, New York: W. W. Norton [1956], p. 104).

22   Eduard Bibring, "The Concept of the Repetition Compulsion", *Psychoanalytic Quarterly* 12 (1943), pp. 75–97.

23   Robert Waelder, "The Psychoanalytic Theory of Play", *Psychoanalytic Quarterly* 2 (1933), pp. 208–24.

24   Alexander, *Psychoanalysis and Psychotherapy*, p. 104.

25   Bibring, *"Concept of the Repetition Compulsion"*.

26   Ernest Hartman, *The Nightmare*, New York: Basic Books Inc. (1984), pp. 194, 243.

27   Roy Grinker and John Spiegel, *Men under Stress*, London, J. & A. Churchill Ltd (1945), p. 348.

28   *Ibid.*

29   Pseudohallucinations are pathognomonic to post-traumatic neurosis. They differ from the psychotic kind, as the person involved is aware that it is not a real perception.

30   Max Stern, *Repetition and Trauma*, Hove and London: Analytic Press (1988), pp. 95–114.

31   Allan Compton, "A Study of the Psychoanalytic Theory of Anxiety", *Journal of the American Psychoanalytic Association* 28, 4 (1980), pp. 739–74.

32   S. Freud, *Psychoanalysis and the War Neuroses*, ed. Ernest Jones, International Psycho-Analytic Press (1921), pp. 1-4. By narcissistic neurosis Freud meant that the libido is attached to the self and not to objects.

33   Sándor Ferenczi, "Two types of War Neurosis", in *Further Contributions to the Theory of Technique of Psychoanalysis*, London: Hogarth Press (1955), p. 141.

34   Freud, *Psychoanalysis and the War Neuroses*, p. 2.

35   Ferenczi, "Two Types of War Neurosis".

36   Ernest Simmel, "Introduction into the War Neuroses", in Ernest Johns (ed.), *Psychoanalysis and War Neuroses*, London: International Psychoanalytic Press (19??), pp. 30–43.

37   Grinker and Spiegel, *Men under Stress*.

38   Abraham Kardiner, *War Stress and Neurotic Illness*, New York and London, Paul B. Hoeber Inc. (1947), pp. 201–2.

39   Sandler *et al.*, "Reflections on Some Relations between Psychoanalytic Concepts and Psychoanalytic Practice", *International Journal of Psychoanalysis* 64 (1983), pp. 35–45.

40   Massud Khan, "The Concept of Commulative Trauma", *Psychoanalytic Study of the Child* 18: (1963), pp. 286–306.

41   Sandler *et al.*, "Reflections on Some Relations", *International Journal of Psychoanalysis* 64 (1983), p. 35.

< NOTES TO PP. 74–80>

42   Leo Rangell, "A Further Attempt to Resolve the Problem of Anxiety", *Journal of the American Psychoanalytic Association* 16 (1968), pp. 371–404.
43   S. Freud, *New Introductory Lectures on Psycho-Analysis* (1939), Standard Edition 22 (1973), pp. 82, 84–95.
44   Rangell, "Problem of Anxiety".
45   Clifford Yorke *et al.*, "A Developmental View of Anxiety", *Psychoanalytic Study of the Child* 31 (1976), pp. 107–35.
46   S. Freud, *Introductory Lectures on Psychoanalysis* (1916–17), Standard Edition 15 (1975), p. 8.
47   Robert Fliess, *The Revival of Interest in the Dream*, New York: International Universities Press (1953), p. 11.
48   Angel Garma, *Psychoanalysis of Dreams*, New York: Jason Aronson (1974), pp. 194–208.
49   Pinchas Noy, "Metapsychology as a Multiple System", *International Review of Psychoanalysis* 4 (1977), pp. 1–12.
50   Waelder, "The Psychoanalytic Theory of Play".
51   Freud, *The Interpretation of Dreams*, p. xxvii: "Insight such as this falls to one's lot but once a lifetime." Another piece of evidence for Freud's evaluation of his *Interpretation of Dreams* comes from his statement in *Revision of the Theory of Dreams* (1933), Standard Edition 22 (1973), p. 7: "The theory of dreams has remained what is most characteristic and peculiar about the young science, something to which there is no counterpart in the rest of our knowledge, a stretch of new country, which has been reclaimed for popular beliefs and mysticism."
52   Freud, *Beyond the Pleasure Principle*, p. 14.
53   Freud, "The Question of a *Weltanschauung*" (1933), Standard Edition 22 (1973), p. 165. This is a citation from Heinrich Heine.
54   Max Stern, "Pavor Nocturnus", *International Journal of Psychoanalysis* 32 (1951), pp. 301–9.

Chapter 7   How *Logos* Arose Mythologically from *Mythos*

1   Freud, *Beyond the Pleasure Principle*, pp. 38–41.
2   Plato, *Symposium*, trans. Benjamin Jowett, Library of Liberal Arts, Indianapolis: Bobbs-Merrill Co. Inc. (1956).
3   Freud, *Beyond the Pleasure Principle*, pp. 36–7.
4   Emanuel Kant, "Idea for a Universal History with Cosmopolitan Intent", prop. 4, in C. J. Friedrich (ed.), *The Philosophy of Kant*, New York: Modern Library (1949), p. 144: "It seems, then, that an instinct is an image inherent in organic life to restore an earlier state of things which the living entity has been obliged to abandon under the pressure of external disturbing forces; that is, it is a kind of organic elasticity, or, to put it in another way, the expression of the inertia in organic life."
5   S. Freud, *Civilization and Its Discontents* (1930), Standard Edition 21 (1973), p. 64. Romain Roland described the oceanic feeling as a sensation of eternity, a feeling of something limitless. Freud's reaction to this was: "I cannot discover this 'oceanic' feeling in myself. It is not easy to deal scientifically with feelings . . . If I understood my friend rightly, he means the same thing

< Notes to pp. 80–89>

by it as the consolation offered by an original and somewhat eccentric dramatist to his hero who is facing a self-inflicted death: 'We cannot fall out of this world.' That is to say, it is a feeling of an indissoluble bond, of being one with the external world as a whole. I may remark that to me this seems something rather in the nature of an intellectual perception ... From my own experience I could not convince myself of the primary nature of such a feeling ... There is nothing of which we are more certain than the feeling of ourself, of our own ego. This ego appears to us as something autonomous and unitary, marked off distinctly from everything else."

6   Freud, *Beyond the Pleasure Principle*, p. 63.
7   *Ibid.*
8   Leo Rauch, *Faith and Revolution: The Philosophy of History*, Tel Aviv: Yahdav, (1978) (Hebrew).
9   Freud, "The Question of a *Weltanschauung*", pp. 158–82.
10   Freud, *Moses and Monotheism*, pp. 7–137.
11   *Ibid.*, pp. 113–15.
12   Freud, *Moses and Monotheism*, p. 130.
13   *Ibid.*, p. 115.
14   *Ibid.*, p. 63.
15   Freud, "The Question of a *Weltanschauung*".
16   S. Freud, *Analysis Terminable and Interminable* (1973), Standard Edition 23 (1973), pp. 216–53.
17   The idea of Empedocles differs from Plato's myth in that he does not talk in terms of closed circles but of sinusoids. He thought of the process of the universe as a continuous, never-ceasing alternation of periods, in which the one or the other of the two fundamental forces gains the upper hand. Both Plato's and Empedocles' versions on the philosophy of history have in common a vision in which linearity is ruled out.
18   Freud, *Beyond the Pleasure Principle*.
19   Oswald Spengler, *The Decline of the West*, 2 vols, London: George Allen & Unwin (1971).
20   Ecclesiastes 12.13.

Part IV   *The Epistemological Split in Psychoanalysis*

Chapter 8   The Dictatorship of the *Logos*

1   Freud, "The Question of a *Weltanschauung*".
2   The verse by Heinrich Heine reads: "Mit Seinen Nachtmützen und Schlafröcken stopft er die Lücken des Weltenbaus" (With his nightcaps and nightgowns he fills in the gaps of the world's structure) (*ibid.*, p. 161). "Its [scientific] endeavour is to arrive at correspondence with reality—that which exists outside of us and independently of us. This correspondence with the real external world we call truth ... Intellect—reason—is among the powers which we may most expect to exercise a unifying influence on men. It may be imagined how impossible human society would be if everyone had his own multiplication table and his own private units of length and weight. Our best hope for the future is that intellect—the scien-

< NOTES TO PP. 89–99>

tific spirit, reason—may in the process of time establish a dictatorship in the mental life of man" (p. 171).

3 Freud, "The Question of a *Weltanschauung*".

4 *Ibid.*

5 *Ibid.*

6 *Ibid.*

7 In Gerald Holton, *Thematic Origins of Scientific Thoughts*, Cambridge, MA: Harvard University Press (1988), p. 129.

8 Freud, "The Question of a *Weltanschauung*".

9 Holton, *Thematic Origins of Scientific Thought*, pp. 99-143.

10 Agassi, *Science and Society*.

11 Fritjof Capra, *The Tao of Physics*, Berkeley: Shambhala (1975).

12 P. W. Bridgman, *The Ways Things Are*, Cambridge, MA: Harvard University Press (1959), p. 6.

13 S. Freud, *New Introductory Lectures on Psycho-Analysis* (1932–6), Standard Edition 22 (1973), p. 80.

14 Wolfgang Pauli, "Naturwissenschaflichte und Erkenntnistheoretische Aspekte der Ideen vom Unbewussten", in *Aufsätze und Vorträge der Physik und Erkennthistheorie*, Braunschweig: Friedrich Vieweg (1961), p. 115. See also K. V. Laurikainen, *Beyond the Atom: The Philosophical Thought of W. Pauli*, Berlin: Springer Verlag (1988).

15 Werner Heisenberg, *Physics and Beyond*, in W. H. Heidcamp (ed.), *The Nature of Life*, Baltimore: University of Maryland Press (1978).

16 Niels Bohr, *Atomic Physics and Human Knowledge*, New York: John Wiley & Son (1958), pp. 13–22.

17 George Klein, "One Psychoanalysis or Two", in *Psychoanalytic Theory: An Exploration of Essentials*, New York: International Universities Press (1976), pp. 41–71.

18 Roy Schafer, *A New Language for Psycho-Analysis*, New Haven and London: Yale University Press (1976), pp. 57–101.

19 Paul Ricoeur, *Freud and Philosophy: An Essay on Interpretation*, New Haven: Yale University Press (1970), pp. 65–158.

20 Jürgen Habermas, *Knowledge and Human Interests*, London: Heineman Educational Books Ltd (1972), pp. 214–45.

21 Overdetermination refers to the fact that the formation of the unconscious can be attributed to a plurality of determining factors. See J. Laplanche and J. B. Pontalis, *The Language of Psychoanalysis*, New York and London: W. W. Norton (1973). This would mean, for instance, that the same dream could be open to different interpretations.

22 Multiple function refers to psychic phenomena necessarily being explained via multiple conceptions and functions. Thus, every psychic act can and must always be viewed in every case as a simultaneous attempted solution of different problems.

23 We find this idea in, for example, Ernst Mayer, *The Growth of Biological Thought*, Cambridge, MA: Belknap Press and Harvard University Press (1987), pp. 829–60. Mayer points out the shortcomings of the classical theories of science based exclusively on philosophy.

< NOTES TO PP. 99–102>

Chapter 9   The Psychoanalysis of Epistemology

1   Freud, "The Question of a *Weltanschauung*".
2   Freud's considering falling in love a sort of a disease is a good example of his wish for autarchy ("Observations on Transference Love" [1915], Standard Edition 12 (1962), pp. 157–71.
3   Freud, "The Question of a *Weltanschauung*", p. 176.
4   Bertrand Russell, *The Conquest of Happiness*, London: George Allen & Unwin (1943).
5   Carl Schorske, *Fin de Siècle Vienna: Politics and Culture*, New York: Vintage Books (1981), p. xvii.
6   Holton, *Thematic Origins of Scientific Thought*, p. 195.
7   Freud, *Introductory Lectures on Psychoanalysis*, p. 285.
8   Lewis Feuer, *Einstein and the Generations of Science*, New York: Basic Books (1974), p. 298: "Human megalomania will have suffered its third and most wounding blow from the psychological research of the present time which seeks to prove to the ego that it is not even master in its own house, but must content itself with scanty information of what is going on unconsciously in its mind."
9   S. Freud, *The Future of an Illusion* (1927), Standard Edition 21 (1973), p. 55: "The transformations of scientific opinion are developments, advances, not revolutions."
10   *Ibid.*
11   S. Freud, *Civilization and Its Discontents* (1930), Standard Edition 21 (1973), p. 65. Freud claims that "there is nothing safer than the self feeling of a person. It appears to us independent and well defined towards the outside world."
12   Martin Wangh, "The Genetic Sources of Freud's Differences with Romain Rolland on the Matter of Religious Feelings", in H. Blum *et al.* (eds), *Fantasy, Myth, Reality*, New York: International Universities Press (1988), p. 259–85. According to Wangh, Freud believes that the human being should save himself from darkness of the infant/mother relationship where all infantile helplessness it to be found. The oceanic feeling is an atavistic remainder of the mother/child dyad. For Freud death is connected to the woman, the goddess of death. Therefore Freud turned from undifferentiated oceanic universum to the protection of the father. This is why Freud finds safety in the intellect, which is represented by the father. Wangh came to his conclusions following an analysis of Freud's relationship with both his parents.
13   Maslow, *Psychology of Science*, pp. 20–33. According to Maslow, science can be a defense. It can be primarily a philosophy of safety, a security system, a way of avoiding anxiety and upsetting problems. In the extreme instance it can be a way of avoiding life, a kind of self-cloistering. It can become a social institution with primarily defensive, conserving functions, ordering and stabilizing rather than discovering and renewing.
14   Sambursky, *Laws of Heaven and Earth*.

< Notes to pp. 106–110>

15 Maslow, *Psychology of Science*, pp. 20–33.
16 Shmuel Hugo Bergman, *Dialogical Philosophy from Kierkegaard to Buber*, Jerusalem: Bialik Institute (1974) (Hebrew).
17 George Devereux, *From Anxiety to Method*, The Hague and Paris: Mouton (1967), p. 287.

Chapter 10   Complementarity in Psychoanalysis

1   Niels Bohr, *Quantum Physics and Philosophy: Essays 1958–1962*, New York: Inter-Science Publishers, 1963, p. 7.
2   Klaus Michael Mayer-Abich, "Komplementarität", in J. Ritter and K. Gründer (eds), *Historisches Wörterbuch der philosophie,* Basel: Schwabe & Co. (1967), vol. 4, pp. 933–4.
3   George Devereux, "Freud, the Discoverer of the Principle of Complementarity", *International Review of Psychoanalysis* 7 (1980), p. 521.
4   Noy, "Metapsychology as a Multiple System", p. 1.
5   Arnold Modell, *Psychoanalysis in a New Context*, New York: International Universities Press (1984), pp. 11–22.
6   *Ibid.*
7   Heinz Kohut, *The Restoration of the Self*, New York: International Universities Press (1977), p. 68.
8   Heinz Kohut, "Introspection, Empathy and Psychoanalysis", *Journal of the American Psychoanalytic Association* 7 (1959), pp. 459–83.
9   Heinz Kohut, *How Does Psychoanalysis Cure?*, Chicago: University of Chicago Press (1984), chap. 4.
10 Schafer, *The Analytic Attitude*, pp. 7–8.
11 Paul Feyerabend, *Farewell to Reason*, London and New York: Verso (1987); *idem, Against Method*, London and New York: Verso (1975), pp. 9–13.
12 Freud, "The Question of a *Weltanschauung*".
13 Efraim Katzir, *In the Crucible of Scientific Revolution*, Tel Aviv: Am Oved (1971) (Hebrew).
14 Henry Edelheit, "Complementarity as a Rule in Psychoanalytic Research", *International Journal of Psychoanalysis* 57 (1976), pp. 23–30.
15 Popper, *Conjectures and Refutations*, p. 231.
16 Günter Stent, *The Coming of the Golden Age*, New York: Natural History Press (1969), p. 18.
17 Alice Miller, *Am Anfang war Erziehung*, Frankfurt: Suhrkamp (1980).
18 Max Delbrück, *Mind from Matter*, Oxford: Blackwell Scientific Publications Inc. (1986), p. 249.
19 Bohr, *Quantum Physics and Philosophy*.
20 Devereux, *From Anxiety to Method*, p. 287.

Chapter 11   Countertransference as a Lacuna in Psychoanalysis

1   Horacio Etchegoyen, *The Fundamentals of Psychoanalytic Technique*, London: Karnac Books (1991), p. 93.
2   Bridgman, *The Way Things Are*, p. 6.
3   Peter Gay, *A Life for Our Time*, New York, W. W. Norton (1988).

< NOTES TO PP. 110–122>

4   Heinz Kohut, "Creativeness, Charisma, Group Psychology: Reflections on the Self-Analysis of Freud", *Psychological Issues* 9, 2/3 (1976), pp. 379–425.

5   S. Freud, "The Future Prospects of Psychoanalytic Therapy" (1910), Standard Edition 11 (1973), pp. 139–51.

6   Freud, "Observations on Transference Love", pp. 152–71.

7   *Ibid.*

8   *Ibid.*, p. 168.

9   Miller, *Am Anfang War Erziehung*.

10 Etchegoyen, *The Fundamentals of Psychoanalytic Technique*, pp. 259–64.

11 James Kern, "Countertransference and Spontaneous Screens", *Journal of the American Psychoanalytic Association* 26 (1978), pp. 21–45.

12 Paula Heimann, "On Countertransference", *International Journal of Psychoanalysis* 31 (1950), pp. 81–4.

13 Heinrich Racker, "A Contribution to the Problem of Countertransference", *International Journal of Psychoanalysis* 34 (1953), pp. 313–24.

14 Margaret Little, "Countertransference and the Patient's Response to It", *International Journal of* Psychoanalysis 32 (1951), pp. 32–40.

15 Devereux, *From Anxiety to Method*, pp. 274–92.

16 Devereux, *Ethnopsychoanalysis*, pp. 1–19.

17 *Ibid.*

18 Devereux, *From Anxiety to Method*, pp. 274–92.

19 *Ibid.*

20 Schafer, *A New Language for Psychoanalysis*, . See, for instance, pp. 22–56.

21 Donald Spence, *Narrative Truth and Historical Truth*, New York: W. W. Norton (1982), pp. 279–98.

22 Eveline Schwaber, "Construction, Reconstruction and the Mode of Clinical Atunement", *International Journal of Psychoanalysis* 60, 4 (1979), pp. 273–91.

23 Arnold Cooper, "Changes in Psychoanalytic Ideas: Transference Interpretation", *Journal of the American Psychoanalytic Association* (1987), pp. 77–98.

24 Little, "Countertransference", pp. 32–40.

25 Donald Winnicott, "Hate in the Countertransference", in *Collected Papers: Through Pediatrics to Psychoanalysis*, New York: Basic Books (1958), pp. 194–203.

26 Lucia Tower, "Countertransference", *Journal of the American Psychoanalytic Association* 4 (1956), pp. 224–65.

27 Joseph Sandler, "Countertransference and Role-Responsiveness", *International Review of Psychoanalysis* 43, 3 (1976), pp. 43–7.

28 James McLaughlin, "Transference, Psychic Reality and Countertransference", *Psychoanalytic Quarterly* 50, 4 (1981), pp. 639–64.

29 Roy Schafer commented on Kohut's epistemology in his "Action Language and the Psychology of the Self", *Annual of Psychoanalysis* 8 (1980), pp. 1–10. According to Schafer, Kohut supposed that direct introspection and empathy are possible, whereas they are themselves to be viewed as constructions and as methods that cannot be theory-free. The model of empathy requires both an empathic analyst and a person to be empathized

< NOTES TO PP. 122–129>

with, as against engaging in some kind of direct and theory-free contact and observation.

30 Barry Protter, "Ways of Knowing in Psychoanalysis", *Contemporary Psychoanalysis* 24, 3 (1988), pp. 498–526.

31 Charles Hanley, "The Concept of Truth in Psychoanalysis", *International Journal of Psychoanalysis* 71, 3 (1990), pp. 375-83.

32 Etchegoyen, *The Fundamentals of Psychoanalytic Technique*, p. 265.

33 Robert Stolorow, "Intersubjectivity Psychoanalytic Knowing and Reality", *Contemporary Psychoanalysis* 24, 2 (1988), pp. 331–8.

Part V Evolution, Revolution, Revelation and Creativity in Psychoanalysis

Chapter 12 Private Theories in Science

1 Sandler: "Reflections on Some Relations between Psychoanalytic Concepts and Psychoanalytic Practice", *International Journal of Psychoanalysis* 64 (1983), p. 35.

2 *Ibid.* Sandler comments: "It is my firm conviction that the investigation of the implicit, private theories of clinical psychoanalysts opened a major new door in psychoanalytic research. One of the difficulties in understanding such research is that posed by the conscious or unconscious conviction of many analysts that they do not do 'proper analysis'."

3 *Ibid.*

4 Karl Popper, *The Open Society and Its Enemies,* Princeton, New Jersey: Princeton University Press (1971), pp. 169–201.

Chapter 13 Leaps versus Gradual Development

1 Marcellin Berthelot, *La Revolution chimique: Lavoisier*, Paris: F. Alcan, 1890.

2 Kuhn, *The Structure of Scientific Revolutions*.

3 Hanson, *Patterns of Discovery*.

4 Arthur Eddington, *On the Nature of the Physical world*, London: Everyman's Library (1947).

5 George Sarton, *A History of Science*, Cambridge, MA: Harvard University Press (1952).

6 Pierre Duhem, *The Aim and Structure of the Physical Theory*, Princeton: Princeton University Press (1991), pp. 239.

7 Herald Hoefding, *The Problem of Philosophy*, New York: G. Fisch (1905).

8 Thomas Kuhn, *The Essential Tension*, Chicago: Chicago University Press (1977), p. 227.

9 Moshe Zukermann, *The Historiography of the French Revolution*, Jerusalem: Defence Ministry Press (1990), pp. 42–3.

10 S. Freud, *A Difficulty in the Path of Psychoanalysis*, Standard Edition 17 (1973), pp. 140–1.

11 Freud, *Analysis Terminable and Interminable*, pp. 216–53.

12 Freud, *Difficulty in the Path*, pp. 143–4.

< Notes to pp. 129–140>

13   S. Freud, *Joseph Popper-Linkeus and the Theory of Dreams* (1923), Standard Edition 19 (1973), p. 261.

14   S. Freud, *A Note on the Pre-History of the Technique of Psychoanalysis*, Standard Edition 18, p. 264.

15   Freud, *The Future of an Ilusion*, p. 55.

16   Freud, "The Question of a *Weltanschauung*".

17   Freud, *The Interpretation of Dreams*.

18   Devereux, *From Anxiety to Method*, p. 287.

19   Sambursky, *The Laws of Heaven and Earth*, p. 56.

20   Foucault, *The Order of Things*, pp. 19–20.

21   Holton, *Thematic Origins of Scientific Thought*, p. 195.

22   Schorske, *Fin de Siècle*, p. 11.

23   Jaacov Golomb, *Between Nietzsche and Freud*, Jerusalem: Magnes University Press.

24   Adolf Grünbaum, *The Foundations of Psychoanalysis: A Philosophical Critique*, Berkeley: University of California Press (1984).

25   Freud, *Difficulty in the Path*, pp. 140–1.

26   Freud, *The Psychopathology of Everyday Life*, Standard Edition 6 (1973).

27   Freud, *Totem and Taboo*.

28   Stephan Mitchel, *Hope and Dread in Psychoanalysis,* New York: Basic Books (1993), pp. 13–91.

29   Ricardo Steiner, "Some Thoughts on Tradition and Change Arising from the Examination of British Society", *International Review of Psychoanalysis* 12 (1985), pp. 12–27.

Epilogue   Scientific Creativity in the Shadows

1   Bush, "Psychoanalysis and Scientific Creativity".

2   Peter L. Giovacchini, "On Scientific Creativity", *Journal of the American Psychoanalytic Association* 8 (1960), pp. 407–26.

3   Ernst Kris, "On Preconscious Mental Processes" (1950), in *Psychoanalytic Explorations in Art*, New York: International Universities Press (1952), pp. 303–18.

4   Laurence Kubie, *Neurotic Distortions: The Creative Process*, New York: Noonday Press (1958).

5   Arthur Koestler, *The Act of Creation*, London: Hutchinson (1964).

6   Jacques Hadamard, *The Psychology of Invention in the Mathematical Field*, Princeton: Princeton University Press (1949).

< Notes to pp. 140–156>

Bibliography

Agassi, Joseph, *Science and Society*, Dordrecht, Boston and London: Reidel Publishing Co., 1981.

Alexander, Franz, *Psychoanalysis and Psychotherapy*, New York: W. W. Norton, 1956.

Anzieu, Didier, *A Skin for Thought*, London: Karnac Books, 1990.

Ariès, Philipe, *The Hour of Our Death*, trans. Helen Weaver, Oxford: Oxford University Press, 1981.

Bachelard, Gaston, *Essai sur la connaissance approchée*, Paris: J. Vrin, 1973.

Bachelard, Gaston, *La Philosophie du Non: Essai d'une philosophie du nouvel esprit scientifique*, Paris: Presses Universitaire de France, 1975.

Bachelard, Gaston, *Le Matérialisme rationel*, Paris: Presses Universitaire de France, 1972.

Bachelard, Gaston, *The New Scientific Spirit*, Boston: Beacon Press, 1984.

Bachelard, Gaston, *The Psychoanalysis of Fire*, Boston: Beacon Press, 1964.

Balint, Enid, "On Being Empty of Oneself", *International Journal of Psychoanalysis* 44 (1963), pp. 470–80.

Barber, Bernard, "Resistance by Scientists to Scientific Discovery", in Bernard Barber and Walter Hirsch (eds), *The Sociology of Science*, New York: Free Press, 1962, pp. 539–56.

Barnes, Barry, *T. S. Kuhn and Social Science*, London: Macmillan, 1982.

Barthley, William, *The Retreat to Commitment*, New York: Alfred Knopf, 1962.

Bergman, Shmuel Hugo, *Dialogical Philosophy from Kierkegaard to Buber*, Jerusalem: Bialik Institute, 1974 (Hebrew).

Berthelot, Marcellin, *La Revolution chimique: Lavoisier*, Paris: F. Alcan, 1890.

Bibring, Eduard, "The Concept of the Repetition Compulsion", *Psychoanalytic Quarterly* 12 (1943), pp. 75–97.

Bick, Esther, "The Experience of the Skin in Early Object-Relations", *International Journal of Psychoanalysis* 49 (1968), pp. 484–6.

Blumenberg, Hans, *Work on Myth*, Cambridge, MA: MIT Press, 1985.

Bohr, Niels, *Atomic Physics and Human Knowledge*, New York: John Wiley & Sons, 1958.

Bohr, Niels, *Quantum Physics and Philosophy: Essays 1958–1962*, New York: Inter-Science Publishers, 1963.

Boring, Edwin, *History, Psychology and Science*, in Robert I. Watson and Donald T. Campbell (eds), New York: John Wiley & Sons, 1963.

< BIBLIOGRAPHY >

Bridgman, Percy Williams, *The Way Things Are*, Cambridge, MA: Harvard University Press, 1959.

Briffault, Robert, *The Mother*, London and New York: George Allen & Unwin, 1959.

Bruner, José, *The Politics of Psychoanalysis*, Oxford: Basil Blackwell, 1995.

Bush, Marshall, "Psychoanalysis and Scientific Creativity", *Journal of the American Psychoanalytic Association* 17 (1969), pp. 136–89.

Capra, Fritjof, *The Tao of Physics*, Berkeley: Shambhala, 1975.

Cassirer, Ernst, *An Essay on Man*, New York: Doubleday, 1973.

Cassirer, Ernst, *The Myth of the State*, New York: Dover Publications, 1946.

Cassirer, Ernst, *The Philosophy of Symbolic Forms*, 3 vols, New Haven and London: Yale University Press, 1955.

Chadwick, Mary, *The Psychological Effects of Menstruation*, Washington, New York: Nervous and Mental Disease Publishing Co., 1932.

Chasseguet-Smirgel, Jannine, "*Blood and Nation*", Panel on the Psychodynamics of Nationalism—Past and Present, Amsterdam: International Congress of Psychoanalysis, 1993

Compton, Allan, "A Study of the Psychoanalytic Theory of Anxiety", *Journal of the American Psychoanalytic Association* 28 (1980), pp. 739–74.

Comte, Auguste, *Introduction to Positive Philosophy*, ed. Frederick Ferré, Indianapolis: Bobbs–Merrill Co. Inc., 1970.

Cooper, Arnold, "Changes in Psychoanalytic Ideas: Transference Interpretation", *Journal of the American Psychoanalytic Association* 35 (1987), pp. 77–98.

Daly, Claude Dagmar, "Der Menstruationskomplex", *Imago* 14 (1928), pp. 11–75.

Daly, Claude Dagmar, "Hindu Mythologie und Kastrationskomplex", *Imago* 13 (1927), pp. 145–98.

Daly, Claude Dagmar, "Zu meinen Arbeiten über die weiblichen Tabuvorschriften", *Zeitschrift für Psychoanalytische Pädagogik* 5–6 (1933), pp. 225–8.

Davidson, Arnold, "How to Do History of Psychoanalysis: A Reading of Freud's *Three Essays on the Theory of Sexuality*", in François Meltzer (ed.), *The Trials of Psychoanalysis*, Chicago: Chicago University Press, 1987, pp. 39–64.

Delbrück Max, *Mind from Matter*, Oxford: Blackwell Scientific Publications Inc., 1986.

Deutsch, Helene, *The Psychology of Women*, vol. 2, New York: Grune & Stratton, 1944.

Devereux, George, *Ethnopsychoanalysis*, Berkeley: University of California Press, 1978.

Devereux, George, "Freud, the Discoverer of the Principle of Complementarity", *International Review of Psychoanalysis* 7 (1980), p. 521.

Devereux, George, *From Anxiety to Method*, The Hague and Paris: Mouton, 1967.

Devereux, George, *A Study of Abortion in Primitive Societies*, New York: International Universities Press, 1973.

Devereux, George, "The Psychology of Feminine Genital Bleeding", *International Journal of Psychoanalysis* 31 (1950), pp. 237–57.

< BIBLIOGRAPHY >

Douglas, Mary, *How Institutions Think*, New York: Syracuse University Press, 1986.

Duhem, Pierre, *The Aim and Structure of Physical Theory*, Princeton, NJ: Princeton University Press, 1991.

Durkheim, Emile, *Incest: The Origin and Nature of the Taboo*, Lyl: Stuart, 1963.

Eddington, Arthur, *On the Nature of the Physical World*, London: Everyman's Library, 1947.

Edelheit, Henry, "Complementarity as a Rule in Psychoanalytic Research", *International Journal of Psychoanalysis* 57 (1976), pp. 23–30.

Ehrenwald Jan, "History of Psychoanalysis", in Jules Masserman (ed.), *Science and Psychoanalysis*, New York: Grune & Stratton, 1958, pp. 148–70.

Eliade, Mircea, *Le Mythe de l'éternel retour: archétypes et répétition*, Edition Gallimard, 1969.

Erikson, Eric, "Womanhood and the Inner Space", in *Identity*, New York: W. W. Norton, 1968.

Etchegoyen, Horacio, *The Fundamentals of Psychoanalytic Technique*, London: Karnac Books, 1991.

Ferenczi, Sándor, *Final Contributions to the Problem and Method of Psychoanalysis*, London: Hogarth Press and Institute of Psychoanalysis, 1955.

Ferenczi, Sándor, "Two Types of War Neurosis", in *Further Contributions to the Theory of Technique of Psychoanalysis*, London: Hogarth Press and Institute of Psychoanalysis, 1955, pp. 141–3.

Feuer Lewis, *Einstein and the Generations of Science*, New York: Basic Books, 1974.

Feyerabend, Paul, *Against Method*, London and New York: Verso, 1988.

Feyerabend, Paul, *Farewell to Reason*, London and New York: Verso, 1987.

Fisher, Charles, "Multiple Induced Abortion", *Journal of the Psychoanalytic Association* 22 (1974), pp. 395–407.

Fisher, Charles, "A Psychophysiological Study of Nightmares and Night Terrors", *Journal of Nervous and Mental Diseases* 2 (1973), pp. 75–97.

Fleck, Ludwig, *Genesis and Develpment of a Scientific Fact*, Chicago: University of Chicago Press, 1970.

Fliess, Robert, *The Revival of Interest in the Dream*, New York: International Universities Press, 1953.

Foucault, Michel, *Histoire de la folie a L'âge Classique*, Editions Gallimard, 1972.

Foucault, Michel, *The Order of Things: An Archaeology of the Human Sciences*, New York: Vintage Books, 1970 (French).

Fox, Robin, *The Red Lamp of Incest*, New York: E. P. Dutton, 1980.

Frazer, James, *The New Golden Bough*, New York: Criterion Books, 1959.

Freeman, Thomas, *A Psychoanalytic Study of Psychoses*, New York: International Universities Press, 1973.

Freud, S., "Analysis Terminable and Interminable" (1937), S.E. 23 (1973), pp. 216–53.

Freud, S., *Beyond the Pleasure Principle* (1920), S.E. 17 (1973).

Freud, S., *Civilization and Its Discontents* (1930), S.E. 21 (1973).

Freud, S., *The Complete Letters of Sigmund Freud*, Cambridge, ed. Jeffrey Mason, Cambridge, MA: Harvard University Press, 1986.

Freud, S., *A Difficulty in the Path of Psychoanalysis* (1917), S.E. 17 (1973).

< BIBLIOGRAPHY >

Freud, S., *The Future of an Illusion* (1927), S.E. 21 (1973).

Freud, S., "The Future Prospect of Psychoanalytic Therapy" (1910), S.E. 11 (1973), pp. 139–51.

Freud, S., *Instincts and Their Vicissitudes* (1915), S.E. 14 (1973).

Freud, S., *The Interpretation of Dreams* (1900), S.E. 4–5 (1973).

Freud, S., *Joseph Popper-Lynkeus and the Theory of Dreams* (1923), S.E. 19 (1973).

Freud, S., *Moses and Monotheism* (1939), S.E. 23 (1973), pp. 7–137.

Freud, S., *New Introductory Lectures on Psycho-Analysis* (1932–6), S.E. 22 (1973).

Freud, S., *A Note on the Pre-History of the Technique of Psychoanalysis*, S.E. 18 (1973).

Freud, S., "Observations on Transference Love" (1915), S.E. 12 (1974), pp. 157–71.

Freud, S., "The Question of a *Weltanschauung*" (1933), S.E.22 (1973).

Freud, S., *Psychoanalysis and the War Neuroses*, ed. Ernest Jones, London: International Psychoanalytic Press, 1921.

Freud, S., *The Psychopathology of Every-Day Life* (1901), S.E. 6 (1973).

Freud, S., "Revision of the Theory of Dreams" (1933), S.E. 22 (1973), pp. 7–30.

Freud, S., *The Taboo of Virginity* (1918), S.E. 11 (1973).

Freud, S., *Totem and Taboo* (1913), S.E. 13 (1962), pp. 1–161.

Freud, S., *The Unconscious* (1915), S.E. 14 (1973).

Fromm-Reichman, Frieda, *Principles of Intensive Psychotherapy*, Chicago: University of Chicago Press, 1950.

Frosch, John, *The Psychotic Process*, New York: International Universities Press, 1983.

Funkenstein, Amos, *Theology and the Scientific Imagination*, Princeton, NJ: Princeton University Press, 1986.

Gay, Peter, *Freud, a Life for Our Time*, New York: W. W. Norton, 1988.

Gell, Alfred, "Reflections on a Cut Finger", in R. H. Hook (ed.), *Phantasy and Symbol*, New York: Academic Press, 1979, pp. 137–41.

Giddens, Anthony, *The Consequences of Modernity*, Stanford, CA: Stanford University Press, 1989.

Gilman, Sander, *The Case of Sigmund Freud*, Baltimore: The Johns Hopkins University Press, 1992.

Giovacchini, Peter, "On Scientific Creativity", *Journal of the American Psychoanalytic Association* 8 (1960), pp. 407–26.

Gleick, James, *Chaos: Making a New Science*, New York: Viking, 1987.

Goedel, Kurt, *On Formally Undecidable Propositions*, New York: Basic Books, 1962.

Goodall, Jane, *In the Shadow of Man*, Boston: Houghton Mifflin, 1971.

Grinker, Roy R. and Spiegel, John J., *Men under Stress*, London: J. & A. Churchill Ltd, 1945.

Grünbaum, Adolf, *The Foundations of Psychoanalysis: A Philosophical Critique*, Berkeley: University of California Press, 1984.

Habermas, Jürgen, *Knowledge and Human Interests*, London: Heinemann Educational Books Ltd, 1972.

Hadamard, Jacques, *The Psychology of Invention in the Mathematical Field*, Princeton, NJ: Princeton University Press, 1949.

Haeckel, Ernst, *Anthropogenie oder Entwicklungsgeschichte des Menschen*, Leipzig: Engelmann, 1874.

< Bibliography >

Hanley, Charles, "The Concept of Truth in Psychoanalysis", *International Journal of Psychoanalysis* 71 (1990), pp. 375–83.

Hanson, Norwood Russell, *Patterns of Discovery*, Cambridge: Cambridge University Press, 1958.

Harris, Marvin, *The Rise of the Anthropological Theory*, London: Routledge & Kegan Paul, 1968.

Hartman, Ernest, *The Nightmare*, New York: Basic Books Inc., 1984.

Hawking Stephen, *A Brief History of Time*, New York: Bantam Books, 1988.

Haynal, André and Falzeder, Ernst, *One Hundred Years of Psychoanalysis: Contributions to the History of Psychoanalysis*, London: Karnac Books, 1995.

Heimann, Paula, "On Countertransference", *International Journal of Psychoanalysis* 31 (1950), pp. 81–4.

Heisenberg, Werner, *Physics and Beyond*, in W. H. Heidcamp (ed.), *The Nature of Life*, Baltimore: University of Maryland Press, 1978.

Hoefding, Herald, *The Problem of Philosophy*, New York: G. Fisch, 1905.

Holton, Gerald, *The Scientific Imagination: Case Studies*, Cambridge: Cambridge University Press, 1978

Holton Gerald, *Thematic Origins of Scientific Thought: Kepler to Einstein*, Cambridge, MA: Harvard University Press, 1973.

Horney, Karen, "The Dread of Women", *International Journal of Psychoanalysis* 13 (1932), pp. 348–60.

Joravski, David, *The Lysenko Affair*, Cambridge, MA: Harvard University Press, 1970.

Kant, Emanuel, *Idea for a Universal History with Cosmopolitan Intent*, prop. 4, in Carl J. Friedrich (ed.), *The Philosophy of Kant*, New York: Modern Library, 1949.

Kardiner, Abraham, *War Stress and Neurotic Illness*, New York and London: Paul B. Hoeber Inc., 1947.

Katzir, Efraim, *In the Crucible of Scientific Revolution*, Tel Aviv: Am Oved, 1971 (Hebrew).

Kern, James, "Countertransference and Spontaneous Screens", *Journal of the American Psychoanalytic Association* 26 (1978), pp. 21–5.

Kerneberg, Otto, *Borderline Conditions and Pathological Narcissism*, New York: Jason Aronsohn, 1975.

Kestenberg, Judith, "Inside and Outside: Male and Female", *Journal of the American Psychoanalytic Association* 16 (1968), pp. 457–519.

Khan, Massud, "The Concept of Commulative Trauma", *Psychoanalytic Study of the Child* 18 (1963), pp. 286–306.

Klein, George, "One Psychoanalysis or Two", in *Psychoanalytic Theory: An Exploration of Essentials*, New York: International Universities Press, 1976.

Klein, Melanie, "The Oedipus Complex in the Light of Early Anxieties", in *Contributions to Psychoanalysis, 1921–1945*, London: Hogarth Press, 1945, pp. 85–115.

Klein, Melanie, *The Psychoanalysis of Children*, London: Hogarth Press, 1932.

Koestler, Arthur, *The Act of Creation*, London: Hutchinson, 1964.

Kohut, Heinz, "Creativeness, Charisma, Group Psychology: Reflections on the Self-Analysis of Freud", *Psychological Issues* 9, 2/3 (1976), pp. 379–425.

< BIBLIOGRAPHY >

Kohut, Heinz, *The Analysis of the Self*, New York: International Universities Press, 1971.

Kohut Heinz, *How Does Psychoanalysis Cure?*, Chicago: University of Chicago Press, 1984.

Kohut, Heinz, "Introspection, Empathy and Psychoanalysis", *Journal of the American Psychoanalytic Association* 7 (1959), pp. 459–83.

Kohut, Heinz, *The Restoration of the Self*, New York: International Universities Press, 1977.

Kris, Ernst, "On Preconscious Mental Processes", in *Psychoanalytic Explorations in Art*, New York: International Universities Press, 1952, pp. 303–18.

Kubie, Laurence, "A Critical Analysis of the Concept of Repetition Compulsion", *International Journal of Psychoanalysis* 20 (1939), pp. 390–402.

Kubie, Laurence, *Neurotic Distortions: The Creative Process*, New York: Noonday Press, 1958.

Kubie, Laurence, "A Reconsideration of Thinking the Dream Process and the Dreams", *Psychoanalytic Quarterly* 35 (1966), pp. 191–8.

Kuhn, Thomas, *The Essential Tension*, Chicago: Chicago University Press, 1977.

Kuhn, Thomas, *The Structure of Scientific Revolutions*, Chicago: University of Chicago Press, 1970.

LaCapra Dominick, *Historical Criticism*, Ithaca: Cornell University Press, 1985.

LaCapra, Dominick, "History and Psychoanalysis", in F. Meltzer (ed.), *The Trials of Psychoanalysis*, Chicago: Chicago University Press, 1987.

LaCapra, Dominick, "Representing the Holocaust", in S. Friedlaender (ed.), *Probing the Limits of Representation: Nazism and the Final Solution*, Cambridge, MA: Harvard University Press, 1992.

Laplanche, J. and Pontalis, J. B., *The Language of Psychoanalysis*, New York and London: W. W. Norton, 1973.

Laudan, Larry, *Progress and Its Problems: Towards a Theory of Scientific Growth*, Berkeley: University of California Press, 1972.

Laurikainen, K. V., *Beyond the Atom: The Philosophical Thought of Wolfgang Pauli*, Berlin: Springer Verlag, 1988.

Lawton, Henry, "Psychohistorical Methodology", in *The Psychohistorian's Handbook*, New York: Psychohistory Press, 1988.

Lester, Eva and Notman, Malkah, "Pregnancy, Development Crisis and Object Relations", *International Journal of Psychoanalysis* 67 (1968), pp. 357–66.

Levi, Zeev, *Structuralis between Method and* Weltanschauung, Hapoalim Press, 1976 (Hebrew).

Lévi-Strauss Claude, *The Elementary Structure of Kinship*, Boston: Beacon Press, 1969

Lévi-Strauss, Claude, *The Savage Mind*, Chicago: Chicago University Press, 1966.

Lévi-Strauss, Claude, *The Unconscious in Culture*, New York: E. P. Dutton, 1974.

Lidz, Ruth and Lidz, Theodor, "Male Menstruation: A Ritual Alternative to the Oedipal Transition", *International Journal of Psychoanalysis* 58 (1977), pp. 17–31.

Lidz, Theodor, "Nightmares and Combat Neurosis", *Psychiatry* 9 (1946), pp. 37–49.

Little Margaret, "Countertransference and the Patient's Response to It", *International Journal of Psychoanalysis* 32 (1951), pp. 32–40.

< BIBLIOGRAPHY >

Loewenberg, Peter, *Decoding the Past*, Berkeley: California University Press, 1983.

Lupton, Mary, *Psychoanalysis and Menstruation*, Urbana: University of Illinois Press, 1993.

Mach, Ernst, *Knowledge and Error*, Boston: Reidel, 1976.

Mack, John, "Dreams and Psychosis", *Journal of the American Psychoanalytic Association* 17 (1969), pp. 206–21.

Mack, John, "Nightmares, Conflict and Ego Development in Childhood", *International Journal of Psychoanalysis* 45 (1965), pp. 403–28.

Mali, Joseph, "Narrative, Myth and History", *Science in Context* 7 (1994), pp. 121–28.

Malinowski, Bronislaw, *Sex and Repression in Savage Society*, New York: Meridian, 1927.

Maslow, Abraham, *Psychology of Science*, New York: Harper & Row, 1966.

Mayer, Ernst, *The Growth of Biological Thought*, Cambridge, MA: Belknap Press and Harvard University Press, 1987.

Mayer-Abich, Klaus Michael, "Komplementarität", in Joachim Ritter (ed.), *Historisches Wörterbuch der Philosophie*, vol. 4, Basel: Schwabe & Co., 1967.

McLaughlin, James, "Transference, Psychic Reality and Countertransference", *Psychoanalytic Quarterly* 50 (1981), pp. 639–64.

Medawar, Peter B., *Induction and Intuition in Scientific Thought*, Philadelphia: American Philosophical Society, 1969.

Miller, Alice, *Am Anfang war Erziehung*, Frankfurt: Suhrkamp, 1980.

Milner, Marion, *The Hand of the Living God*, London: Hogarth Press, 1969.

Mitchel, Stephen, *Hope and Dread in Psychoanalysis*, New York: Basic Books, 1993.

Modell, Arnold, *Psychoanalysis in a New Context*, New York: International Universities Press, 1984.

Murdoch, George Peter, *Social Structure*, New York: Macmillan, 1949.

Neumann, Erich, *The Great Mother*, Princeton, NJ: Princeton University Press, 1963.

Newell, Fisher, "Multiple Induced Abortion", *Journal of the American Psychoanalytic Association* 22 (1974), pp. 395–407.

Parsons, Talcott, "The Incest Taboo in Relation to Social Structure and the Socialization of the Child", *British Journal of Sociology* 5 (1954), pp. 101–7.

Pauli, Wolfgang, "Naturwissenschaftliche und Erkenntnistheoretische Aspekte der Ideen vom Unbewussten", *Aufsätze und Vorträge der Physik und Erkenntnistheorie*, Braunschweig: Friedrich Vieweg, 1961.

Pinchas, Noy, "Metapsychology as a Multiple System", *International Review of Psychoanalysis* 4 (1977), pp. 1–4.

Pines, Dinora, "Skin Communications, Early Skin Disorders and Their Effect on Transference and Counter-Transference", *International Journal of Psychoanalysis* 61 (1980), pp. 315–22.

Plato, *Symposium*, trans. Benjamin Jowett, Library of Liberal Arts, Indianopolis: Bobbs–Merrill Co. Inc. (1956).

Polanyi, Michael, *Personal Knowledge*, London: Routledge & Kegan Paul, 1962.

Pontalis, Jean Bertrand, *Frontiers in Psychoanalysis*, New York: International Universities Press, 1977.

< BIBLIOGRAPHY >

Popper, Karl, *Conjectures and Refutations*, London: Routledge and Kegan Paul, 1963.

Popper, Karl, *The Logic of Scientific Discovery*, New York: Harper & Row, 1968.

Popper, Karl, *Objective Knowledge*, Oxford: Clarendon Press, 1972.

Popper Karl, *The Open Society and Its Enemies*, Princeton, NJ: Princeton University Press, 1971.

Popper, Karl, *Realism and the Aim of Science*, London: Hutchinson, 1956.

Protter, Barry, "Ways of Knowing in Psychoanalysis", *Contemporary Psychoanalysis* 24 (1988), pp. 498–526.

Racker, Heinrich, "A Contribution to the Problem of Countertransference", *International Journal of Psychoanalysis* 34 (1953), pp. 313–24.

Rangell, Leo, "A Further Attempt to Resolve the Problem of Anxiety", *Journal of the American Psychoanalytic Association* 16 (1968), pp. 371–404.

Raphael-Leff, Joan, *Pregnancy: The Inside Story*, London: Sheldon Press, 1993.

Raphael-Leff, Joan, *Psychological Processes of Childbearing*, London: Chapmant & Hall, 1991.

Rauch, Leo, *Faith and Revolution: The Philosophy of History*, Tel Aviv: Yachdav Press, 1978 (Hebrew).

Renik, Owen, "An Example of Disavowal Involving the Menstrual Cycle", *Psychoanalytic Quarterly* 53 (1984), pp. 523–32.

Ricoeur, Paul, *Freud and Philosophy: An Essay on Interpretation*, New Haven: Yale University Press, 1970.

Ritvo, Lucile, *Darvin's Influence on Freud*, New Haven: Yale University Press, 1990.

Rosenfeld, David, *The Psychotic*, London: Karnac Books, 1992.

Rosenfeld Herbert, *Psychotic States*, New York: International Universities Press, 1973.

Rossi, Ino, *The Unconscious in* Culture, New York, E. P. Dutton, 1974.

Rouse, Joseph, *Knowledge and Power: Toward a Political Philosophy of Science*, Ithaca: Cornell University Press, 1987.

Russell, Bertrand, *The Conquest of Happiness*, London: GeorgeAllen & Unwin, 1943.

Sachs Oliver, "Scotoma: Forgetting and Neglect in Science", in Robert B. Silvers (ed.), *Hidden Histories of Science*, London: Granta Books, 1995, pp. 141–88.

Sambursky, Shmuel *et al.*, *The Laws of Heaven and Earth*, Bialik Institute, 1987 (Hebrew).

Sandler, Joseph, "Counter Transference and Role-Responsiveness", *International Review of Psychoanalysis* 3 (1976), pp. 43–7.

Sandler, Joseph, "Reflections on Some Relations between Psychoanalytic Concepts and Psychoanalytic Practice", *International Journal of Psychoanalysis* 64 (1983), pp. 35–45.

Sarton George, *A History of Science*, Cambridge, MA: Harvard University Press, 1952

Saul, Leon, "Psychological Factors in Combat Fatigue", *Psychosomatic Medicine* 7 (1945), pp. 257–72.

Schafer, Roy, "Action Language and the Psychology of the Self", *Annual of Psychoanalysis* 7 (1980), pp. 1–10.

< BIBLIOGRAPHY >

Schafer, Roy, *The Analytic Attitude*, New York: Basic Books, 1983.

Schafer, Roy, *A New Language for Psycho-Analysis*, New Haven and London: Yale University Press, 1976.

Schelling, Joseph, "Myth", *The Encyclopedia of Philosophy*, vol. 5, London: Collier-Macmillan, 1974.

Schilder, Paul, "What Do We Know about the Interior of the Body?", *International Journal of Psychoanalysis* 16 (1935), pp. 355–60.

Schmideberg, Mellita, *Psychoanalytisches zu Menstruation*, Centrale Zeitung für Psychoanalytische Paedagogik, 5/6 (1931) , pp. 190–203.

Schorske, Carl, *Fin de Siècle Vienna: Politics and Culture*, New York: Vintage Books, 1981.

Schwaber, Eveline, "Construction, Reconstruction and the Mode of Clinical Atunement", *International Journal of Psychoanalysis* 60 (1979), pp. 273–91.

Searls, Harold, *Collected Papers on Schizophrenia and Related Subjects*, London: Hogarth Press, 1965.

Seligman, Brenda, "The Incest Barrier", *British Journal of Psychology* 22, pp. 250–276.

Shapin, Steven, *A Social History of Science*, Chicago: Chicago University Press, 1994.

Shavit, Yaakov, *Judaism in the Greek Mirror and the Emergence of the Modern Hellenized Jew*, Tel Aviv: Am Oved, 1992 (Hebrew).

Shepher, Joseph, *Incest: A Biosocial View*, New York and London: Academic Press, 1983.

Shur, Max, *Freud: Living and Dying*, New York: International Universities Press, 1972.

Simmel, Ernst, "Introduction into the War Neuroses", in Ernest Johns (ed.), *Psychoanalysis and War Neuroses*, London: International Psychoanalytic Press, pp. 30–43.

Spence, Donald, *Narrative Truth and Historical Truth*, New York: W. W. Norton, 1982.

Spengler, Oswald, *The Decline of the West*, 2 vols, London: George Allen & Unwin, 1971.

Spitz, René, "The Primal Cavity", *Psychoanalytic Studies of the Child* 10 (1955), pp. 215–241.

Stein, Martin, "States of Consciousness in the Analytic Situation Including a Note on the Traumatic Dream", in M. Schur (ed.), *Drives, Affect and Behavior*, New York: International Universities Press, 1965, vol. 2, pp. 61–86.

Stein, Yehoyakim, "Some Reflections on the Inner Space and Its Contents", *Psychoanalytic Study of the Child* 43 (1988), pp. 291–304.

Steiner, Ricardo, "Some Thoughts about Tradition and Change Arising from the Examination of British Society", *International Review of Psychoanalysis* 12 (1985), p. 27.

Stent, Günter, *The Coming of the Golden Age*, New York: Natural History Press, 1969.

Stern, Max, "Pavor Nocturnus", *International Journal of Psychoanalysis* 32 (1951), pp. 301–9.

Stern, Max, *Repetition and Trauma*, Hove and London: Analytic Press, 1988.

< BIBLIOGRAPHY >

Steward, Howard, "Changes of Inner Space", *International Journal of Psychoanalysis* 66 (1985), pp. 255–64.

Stolorow, Robert, "Intersubjectivity and Psychoanalytic Knowing of Reality", *Contemporary Psychoanalysis* 24, 2 (1988), pp. 331–8.

Sulloway, Frank J., *Freud: Biologist of the Mind*, New York: Basic Books, 1979.

Taylor, Edmund, "On a Method of Investigating the Development of Institution Applied to Laws of Marriage and Descent", *Journal of the Royal Anthropological Institute* 18 (1988), pp. 245–72.

Tidd, Charles, "Symposium on Psychoanalysis and Ethology", *International Journal of Psychoanalysis* 41 (1960).

Tiles, Mary, *Bachelard: Science and Objectivity*, Cambridge: Cambridge University Press, 1984.

Tower, Lucia, "Countertransference", *Journal of the American Psychoanalytic Association* 4 (1956), pp. 224–65.

Vico, Giambattista, *The New Science*, Ithaca: Cornell University Press, 1991.

Vita, Judith, "Amalgamating with the Existing Body of Knowledge", in André Haynal and Ernst Falzeder, *One Hundred Years of Psychoanalysis: Contributions to the History of Psychoanalysis*, London: Karnac Books, 1995.

Vovell, Michel, *Ideologies and Mentalities*, Cambridge: Polity Press, 1987.

Waelder, Robert, "The Psychoanalytic Theory of Play", *Psychoanalytic Quarterly* 2 (1933), pp. 208–24.

Wangh, Martin, "The Genetic Sources of Freud's Differences with Romain Rolland on the Matter of Religious Feelings", in Harold P. Blum *et al.* (eds.), *Fantasy, Myth and Reality*, New York: International Universities Press, 1988, pp. 259–85.

Westermark, Eduard, *The History of Human Marriage*, London: Macmillan, 1921.

White, Hayden, *The Content of the Form*, Baltimore: The Johns Hopkins University Press, 1987.

Winnicott, Donald Woods, "Ego Integration and Child Development", in *The Maturational Process and the Facilitating Environment*, New York: International Universities Press, 1965, pp. 57–58.

Winnicott, Donald Woods, "Hate in the Countertransference", in *Collected Papers: Through Paediatrics to Psycho-analysis*, New York: Basic Books, 1958, pp. 194–203.

Yorke, Clifford *et al.*, "A Developmental View of Anxiety", *Psychoanalytic Study of the Child* 31 (1976), pp. 107–35.

Zuckermann, Moshe, *Historians and the French Revolution*, Jerusalem: Ministry of Defence Press, 1990 (Hebrew).

< BIBLIOGRAPHY >

Index

< INDEX >

< INDEX >

< INDEX >

< INDEX >

< INDEX >